Geology

Wiley Self-Teaching Guides teach practical skills from accounting to astronomy, management to mathematics. Look for them at your local bookstore.

Geology

A Self-Teaching Guide

Barbara W. Murck

John Wiley & Sons, Inc.

New York • Chichester • Weinheim • Brisbane • Toronto • Singapore

This book is printed on acid-free paper. ♾

Published by John Wiley & Sons, Inc.
Published simultaneously in Canada

Library of Congress Cataloging-in-Publication Data:
Murck, Barbara Winifred
Geology : a self-teaching guide / Barbara W. Murck.
p. cm. — (Wiley self-teaching guide series)
Includes index.
ISBN 0-471-38590-5 (acid-free paper : paper)
1. Geology—Study and teaching (Secondary) I. Title.
II. Wiley self-teaching guides.

QE40.M87 2001
550.7'7—dc21 2001024231

Printed in the United States of America
10 9 8 7 6 5 4 3 2

Contents

Acknowledgments

I wish to acknowledge, first and foremost, the extensive contributions to this book made by Brian Skinner, my friend and coauthor on other Wiley projects. Brian selected the photographs—including several great shots from his own extensive collection—and wrote the captions for them. Much more importantly, the thoughts and words he has shared with me on so many other occasions are woven throughout this book; without them, it would not exist.

Many thanks to Jeff Golick of John Wiley & Sons for initiating this project and seeing it through to completion, for providing me with ever-patient and levelheaded counsel, and for having the energy and foresight to forge full speed into our next collaboration. Thanks, too, to Marjorie Graham, Jill Tatara, and Anna Melhorn, who went beyond the call of duty to help in various ways with the photo selections and artwork, and Sibylle Kazeroid for her guidance through editing and production. My daughter Eliza King earned her keep as an editorial assistant, and Riley and Jack helped out with their usual tolerance of such activities. And Cliff, thanks for recommending me, even though you knew it might distract me from other Wiley projects!

Acknowledgments

A Note to the Reader

Welcome to the study of our home planet, Earth. In this book you will learn about the fascinating geologic processes that make the Earth the special place it is. You will find out how our understanding of this planet has changed dramatically over the past few decades as a result of new theories and new technologies for Earth observation. Yet we continue, as well, to build upon the scientific foundations laid by geologists of the past 200 years. Earth is a fascinating planet, and geology—the scientific study of the Earth—is a multifaceted discipline.

If you have purchased this book, you probably fall into one of three categories:

- you studied geology a number of years ago in college or university and want to update your understanding of the subject;
- you are currently studying geology in college or university and want to supplement your understanding of the material covered in lectures and labs;
- you have developed an interest in geology, perhaps through watching shows on public television or hearing about natural events such as earthquakes and volcanic eruptions, and you want to learn more about this fascinating science.

Whatever your background, I hope this book will meet your needs and that you will enjoy reading it and learning more about our planet in the process.

All of the chapters begin with a brief list of the main things you will be learning about. In most of the chapters you will find some text that is set apart like this:

The symbol indicates that this is something special that you might want to think about or an activity or experiment you might want to try for yourself.

Throughout the book you will find questions and answers, as well as a Self-Test at the end of each chapter. Use these to test your comprehension and retention of concepts and vocabulary as you read. The chapters of the book can be read independently of one another. However, each chapter assumes that you have learned the vocabulary and, in some cases, the concepts presented in the preceding chapters. In particular, the **bold** vocabulary terms are only introduced once; they are not redefined in each chapter in which they appear. For this reason, it will probably make the most sense for you to read the chapters in the order in which they are presented.

This book is mainly about physical geology, meaning that it focuses primarily on Earth materials (rocks, minerals, lavas, etc.) and processes (erosion, sedimentation, volcanism, etc.). However, it is difficult to separate physical geology from historical geology—the study of Earth history as preserved in the rock record and of organisms that lived long ago. There are two chapters (chapter 3 and chapter 10) in the book that focus specifically on historical geology. You will also discover that many other scientific disciplines—including astronomy, chemistry, biology, physics, oceanography, hydrology, and meteorology—contribute to our understanding of this complex planet we inhabit. I hope you will enjoy your exploration of planet Earth!

1 Plate Tectonics: A Revolution in Geology

This we know. The Earth does not belong to us; we belong to the Earth.

—attributed to Chief Seattle

It is my opinion that the Earth is very noble and admirable . . . and if it had continued an immense globe of crystal, wherein nothing had ever changed, I should have esteemed it a wretched lump of no benefit to the Universe.

—Galileo Galilei

OBJECTIVES

In this chapter you will learn

- what physical and historical geology are and the differences between them;
- how the theory of plate tectonics revolutionized geology;
- how scientists gathered evidence to support the theory of plate tectonics;
- how internal processes shape the surface of the Earth and make it a dynamic place to live.

1 UNDERSTANDING THE EARTH

Geology is the scientific study of the Earth. Geology is a young science; it has existed as a modern scientific discipline for just over 200 years. The study of the Earth is traditionally divided into two broad subject areas: physical geology and historical geology. **Physical geology** concerns the processes that operate at or beneath the surface of the Earth and the materials on which those processes operate. Some examples of geologic processes are mountain building, volcanic eruptions, river flooding, earthquakes, and the formation of ore deposits. Some examples of geologic materials are minerals, rocks, soils, lava, and water.

Historical geology concerns geologic events that occurred in the past. These events can be read from the rock record. Historical geologists try to answer questions such as when the oceans formed, why the dinosaurs died out, when the Rocky Mountains rose, and when the first trees appeared. Historical geology helps us establish a chronology of events in Earth history and gives us a context for understanding our present-day environment.

There are many more specialized areas of study within the traditional domains of physical and historical geology. For example, volcanologists study volcanoes and eruptions; seismologists study earthquakes; mineralogists study minerals and crystals; paleontologists study fossils and the history of life on the Earth; structural geologists study how rocks break and bend; economic geologists study the formation and occurrence of valuable ore deposits. This specialization is needed because geology encompasses such a broad range of topics.

Geologists are scientists who make a career out of the scientific study of the Earth. Yet to a certain extent we are all geologists. Everyone living on this planet relies on resources from the Earth: water, soil, building stones, metals, fossil fuels, gems, plastics (made from petroleum), ceramics (made from clay minerals), salt (the mineral halite), and many others. We are affected by geologic processes every single day we spend on the surface of this dynamic planet. By learning as much as we can about these processes, we can become better-informed, more responsible caretakers of our home planet.

Name three examples of geologic processes. Try to think of at least one example that was not mentioned in the text.

____

____

Answer: Examples in the text are mountain building, volcanic eruptions, river flooding, earthquakes, and the formation of ore deposits. Some other examples are groundwater movement, oil and coal formation, evaporation, and erosion. Can you think of any more?

Name three examples of geologic materials. Try to think of at least one example that was not mentioned in the text.

____

____

Answer: Examples in the text are minerals, rocks, lava, and water. Some other examples are soil, magma, glacial ice, and natural gas. Can you think of any more?

2 GEOLOGY THEN AND NOW: A SCIENTIFIC REVOLUTION

Even a science as young as geology can have a revolution, and that is what happened in the 1960s. At that time, a brand-new theory emerged and completely changed our understanding of geologic processes. The tools, the methods, and even the language of geology changed as a result of that scientific revolution. If you studied geology prior to the 1960s, you may remember some terms that are no longer in use today. Terms such as "eugeosyncline" and "miogeosyncline" were used to describe topographic features of the Earth's surface that geologists observed but could not explain. With the advent of the theory of plate tectonics, these features took on new meaning. Consequently, geologists began using new terms to describe them. This book will help you learn the vocabulary we use to describe our current understanding of the Earth.

This first chapter—and, indeed, much of the rest of this book—concerns the plate tectonic revolution and how it has informed and transformed our understanding of the Earth. But geology is currently undergoing another, more subtle revolution. This revolution is driven by the ability of scientists to observe and collect information about the Earth as a whole planet, using instruments mounted on satellites. This ability is quite new; remember that no one had ever seen a picture of the whole Earth until the 1960s, when the first photograph was taken of Earth from space.

Satellite images and data collected from outer space provide a scientific foundation for our study of the Earth as an integrated system. **Earth system science,** as this approach is called, is not new in philosophy, but its tools and techniques are very new. These tools are used in a wide range of applications, from weather forecasting to the monitoring of changes in sedimentation rates, measuring the flow of polar ice, locating mineral resources, documenting the extent of oil spills, tracking depletion of stratospheric ozone, and many others. Through Earth system science, geologists are contributing to our understanding of the Earth as a whole, how the Earth changes over time, and the impacts of human actions on the Earth system.

Name at least three applications of Earth system science.

____

____

____

Answer: Examples in the text are weather forecasting, monitoring changes in sedimentation rates, measuring the flow of polar ice, locating mineral resources, documenting the extent of oil spills, and tracking the depletion of stratospheric ozone. Can you think of any more?

3 GEOLOGY BEFORE PLATE TECTONICS

During the 1800s, people favored the idea that the Earth, originally a molten mass, had been cooling and contracting for centuries. Scientists argued that mountain ranges full of folded rocks were expressions of the contraction and shrinkage of the Earth's interior (if the crust didn't contract as much as the interior, it would fold and crumple like the wrinkled skin of a dried prune). Contraction did appear to explain some features of the Earth's surface, but it could not explain the shapes and positions of the continents. Nor did it explain features like great rift valleys, clearly caused by stretching rather than by contraction.

At the beginning of the twentieth century, scientists discovered that the Earth's interior is heated by the decay of naturally occurring radioactive elements. This suggested that the Earth might not be cooling but rather heating up, and therefore expanding. A smaller Earth might once have been covered mostly by continents. As the Earth expanded, the continents would crack into fragments, and eventually the cracks would grow into oceans. The expanding Earth hypothesis did explain the apparent fit between the coastlines of Africa and South America, which look as if they have been ripped apart from each other. But there are other features that this hypothesis did not easily account for, such as folded mountain ranges formed by compression.

To get around the flaws in the expansion and contraction hypotheses, geologists began to search for other ways of explaining the shapes and positions of the continents, oceans, and mountain chains. By the middle of this century, all reasonable suggestions seemed to have been exhausted; the time was ripe for a totally new approach. This approach turned out to be **plate tectonics**—the theory that the continents are carried along on huge slabs, or **plates,** of the Earth's outermost layer. In some places the plates are slowly colliding, forming compressional features like huge mountain ranges. In other places the plates are moving apart, forming expansional features like great rift valleys. The theory of plate tectonics provided, for the first time, a coherent, unified explanation for *all* of these features of the Earth's surface.

What was wrong with the "contracting Earth" hypothesis?

___

___

Answer: It did not adequately explain the shapes and positions of the continents, nor did it explain features like great rift valleys, which appear to have been caused by stretching.

What was wrong with the "expanding Earth" hypothesis?

___

___

Answer: It did not adequately explain features such as folded mountain ranges formed by compression.

4 CONTINENTAL DRIFT AND THE STORY OF WEGENER

This chapter tells the story of how the theory of plate tectonics was conceived and developed and eventually came to be accepted. The modern part of the story began in the early 1900s with a German meteorologist named Alfred Wegener, who had some controversial ideas about the shapes and positions of the continents.

In 1910, Wegener began lecturing and writing scientific papers about continental drift. His **continental drift** hypothesis suggested that the continents have not always been in their present locations but instead have "drifted" and changed positions. Wegener's idea was that the continents had once been joined together in a single "supercontinent," which he called **Pangaea** (pronounced "pan-JEE-ah"), from Greek words meaning "all lands." He suggested that Pangaea had split into fragments like pieces of ice floating on a pond and that the continental fragments had slowly drifted to their present locations.

Wegener presented a great deal of evidence in support of the continental drift hypothesis. Nevertheless, his proposal created a storm of protest in the international scientific community. Part of the problem was that geologists simply could not envision how the continents could move around. Another part of the problem was that geologists had to be convinced that the evidence that the continents had once been joined was truly conclusive. Let's look at some of the evidence for continental drift, so you can judge for yourself. Notice that no single piece of evidence is conclusive on its own. It took several decades and the weight of all this evidence (and more) to finally convince geologists that continental drift really happens.

What was the name of the "supercontinent" proposed by Alfred Wegener?

Answer: Pangaea.

5 EVIDENCE FROM COASTLINES

Look at a map of the world. The Atlantic coastlines of Africa and South America seem to match, almost like puzzle pieces. The southern coast of Australia similarly seems to match part of the coast of Antarctica, and the same is true of some other continental coastlines. Is this apparent fit an accident, or does it support the hypothesis that the continents were once joined together?

To answer the question of whether continents were once joined, we must first recognize that the edge of the land—that is, the shoreline—usually isn't the true edge of the continent. To find the true edge of a continent, we need to locate the place where the rocks of the continent—mostly made of granite—meet the rocks of the ocean floor—mostly made of basalt. (You will learn more about these two important rock types in chapter 5.)

Along a noncliffed shoreline, such as the Atlantic coasts of North America, South America, and Africa, the land usually slopes very gently toward the sea (Figure 1.1). This gently sloping land is called the **continental shelf.** At the edge of the continental shelf there is a sharp drop-off to the steeper **continental slope.** At the bottom of the steep continental slope, the land begins to level off again; this is the **continental rise,** which marks the transition to the flat ocean floor, the **abyssal plain.** The actual place where the granitic rocks of the continent meet the basaltic rocks of the ocean floor is usually covered by sand, mud, and other loose rock particles. The actual shape of the shoreline depends on sea level, the presence or absence of cliffs, and the details of the topography of the continental shelf in any particular locality. Thus, the actual transition from continent to ocean may (or may not) be underwater.

So, how do we identify the true edge of a continent? Usually the edge of a continent is defined as being halfway down the steep continental slope. When we try to fit the continents together, we fit them along this line rather than along the present-day coastline. When we fit Africa and South America together in this way, the result is remarkable (Figure 1.2). In the "best-fit" position, the average gap or overlap between the two continents is only 90 kilometers (km) (about 56 miles

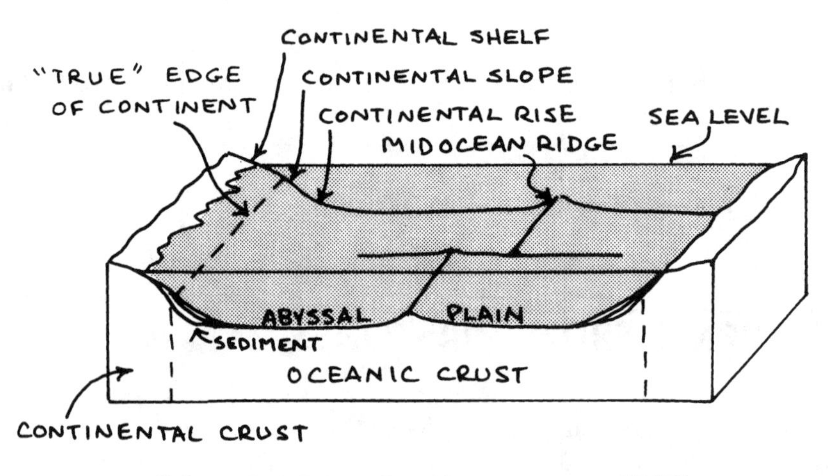

Figure 1.1

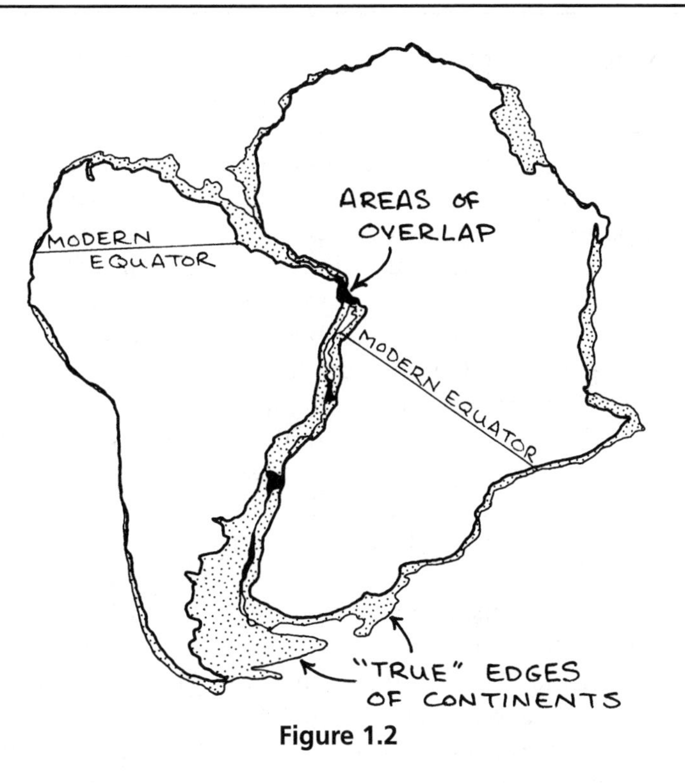

Figure 1.2

[mi]). (Note that 1 kilometer ≈ 0.62 miles; see Appendix 1 for more about units, conversions, and abbreviations.) Furthermore, the most significant overlapping areas consist of rocks that were formed *after* the time when the continents are thought to have split apart. This strongly suggests that Africa and South America were once joined.

Sketch and neatly label a diagram showing the transition from continent to ocean. Show how the slope of the land changes, and label all of the topographic features. On your diagram, indicate the "true" edge of the continent.

Answer: Refer to Figure 1.1.

6 EVIDENCE FROM ROCKS

If Africa and South America were once joined, one would expect to find similar geologic features on both sides of the join. Such correlations provided some of the most compelling evidence presented by Wegener in support of the continental drift hypothesis. However, matching the geology of rocks on opposite sides of an ocean is more difficult than you might imagine. Rock-forming processes never cease. Some rocks formed before the continents were joined, some while they were joined, others during the splitting of the continents, and still others after they separated. How can we tell which rocks are significant in trying to find a match between the continents?

A logical starting point is to see if the ages and orientations of similar rock types match up across the ocean. In Wegener's time, geologists did not have sophisticated tools for determining the exact age of a rock. But now we do have such tools, and we know that there are strong similarities in the ages of rocks across the oceans. The match is particularly good between rocks about 550 million years old and older in northeast Brazil and West Africa, but there is not a good match for younger rocks. This suggests that the two continents were joined together for some period of time prior to 550 million years ago, and they subsequently split apart.

We can also look for continuity of geologic features such as mountain chains. If we rejoin the continents as they would have been in the supercontinent Pangaea, mountain belts of similar ages seem to line up. For example, the oldest portions of the Appalachian Mountains, extending from the northeastern part of the United States through eastern Canada, match up with the Caledonides of Ireland, Britain, Greenland, and Scandinavia. A younger part of the Appalachians lines up with a mountain belt of similar age in Africa and Europe. These and other bedrock features that match up across the oceans are strong evidence that the continents were once joined together.

Another geologic feature that matches across continental joins is the deposits left by ancient ice sheets. These are similar to deposits left by recent glaciers in

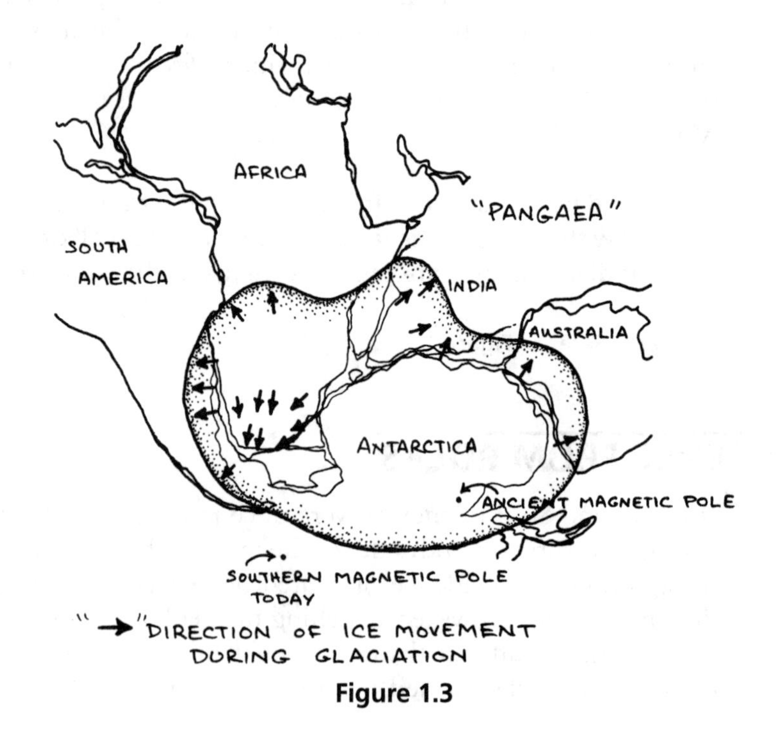

Figure 1.3

Canada, Scandinavia, and the northern United States. In South America and Africa there are very thick glacial deposits. The deposits are the same age, and they match almost exactly when the continents are "moved back together." As glacial ice moves, it cuts grooves and scratches in underlying rocks and produces folds and wrinkles in soft sediments. Such features provide evidence of the direction the ice was moving during the glaciation. When Africa and South America are moved back together, the grooves and scratches show that the ice was radiating outward from the center of a former ice sheet (Figure 1.3). It's hard to imagine how such similar glacial features could have been created if the continents had not once been joined together. Africa and South America must also have had similar climates during this period, colder than their present-day climates. This suggests that they were not in their present equatorial locations. In fact, the southern portion of Pangaea was most likely close to what was then the South Pole.

How can the ages of rocks provide evidence that two continents—now separated from each other by an ocean—were once joined?

____

____

____

Answer: If the continents were once joined, we would expect to find rocks of similar type and age on either side of the ocean. There are strong similarities in rocks about 550 million years old and older in northeast Brazil and West Africa. This suggests that the two continents were joined together for some period of time prior to 550 million years ago.

7 EVIDENCE FROM FOSSILS

If Africa and South America were joined at one time, with the same climate and matching geologic features, then they also should have hosted similar plants and animals. To check this, Wegener turned to the fossil record. This revealed that there were communities of plants and animals that appear to have evolved together until the time of the splitting apart of Pangaea, after which they evolved separately.

Wegener pointed to specific fossils found in matching areas across the oceans. One example he used was an ancient fern, *Glossopteris,* whose fossilized remains have been found in southern Africa, South America, Australia, India, and Antarctica. Could the seeds of this plant have been carried from one location to another across the oceans? Probably not. The seeds of *Glossopteris* were large and heavy, and could not have been carried very far by wind or water currents. This fern flourished in a cold climate; it would not have thrived in the warm present-day climates of the continents where its fossil remains are found. This, too, suggests that the continents once had similar, colder climates.

There are other examples as well. The fossilized remains of *Mesosaurus,* a small reptile, are found both in southern Brazil and in South Africa. The types of rocks in which the fossils are found are very similar. *Mesosaurus* did swim, but was probably too small to swim all the way across the ocean. Fossilized remains of specific types of earthworms also occur in areas that are now widely separated. How could they possibly have migrated across the oceans? The landmasses in which they lived must once have been connected.

How did Wegener use *Glossopteris* to support the hypothesis of continental drift?

____

____

____

Answer: Fossils of *Glossopteris,* an ancient fern, have been found in similar rocks in southern Africa, South America, Australia, India, and Antarctica—locations that are now widely separated by oceans. Its seeds were large and heavy and could not have been transported very far by wind or water. This suggests that the areas where *Glossopteris* fossils are now found must once have been joined together.

8 THE MISSING CLUE: PALEOMAGNETISM

Wegener and his supporters gathered more and more evidence in support of continental drift, but many scientists remained unconvinced. Wegener died in 1930 without seeing the end of the debate, which continued after his death. A turning point occurred in the 1950s through the study of **paleomagnetism,** ancient magnetism preserved in rocks. When lava cools and solidifies into rock, it becomes magnetized and takes on the **polarity**—the north-south directionality—of the Earth's magnetic field at that time. Just as a free-swinging magnet today will point toward today's magnetic north pole, so too, does a rock's paleomagnetism act as a pointer toward the location of the Earth's magnetic north pole at the time of rock formation.

In the 1950s, geologists studying the paleomagnetism of rocks from different localities found evidence suggesting that the Earth's magnetic poles had wandered all over the globe for at least the past several hundred million years. They plotted the pathways of the poles on maps, and referred to the phenomenon as **apparent polar wandering** (Figure 1.4). Geologists were puzzled by this evidence. They knew that it was extremely unlikely that the magnetic poles themselves had moved. Instead they concluded, somewhat reluctantly, that it must have been the continents themselves that had moved, carrying their magnetic rocks with them.

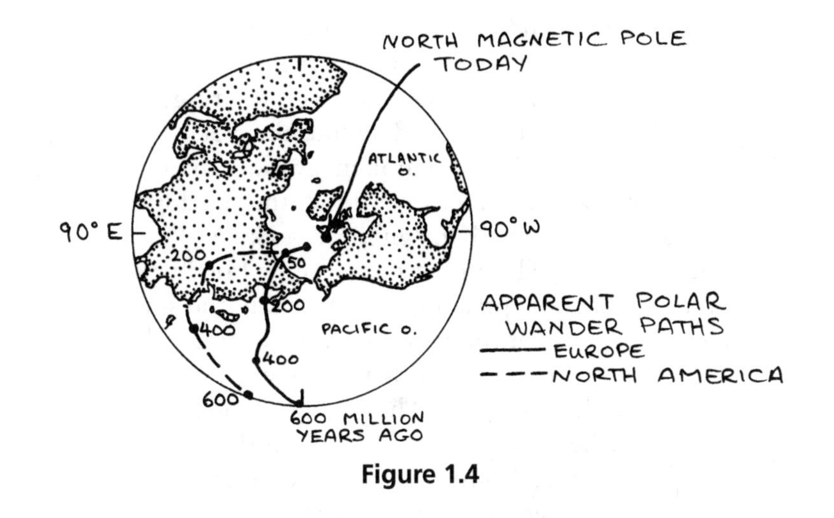

Figure 1.4

What is polarity?

Answer: The north-south directionality of the Earth's magnetic field.

9 SEAFLOOR SPREADING

The new evidence from paleomagnetism helped revive the hypothesis of continental drift. But many scientists were still holding out for a final piece of evidence that would demonstrate conclusively that a supercontinent had actually split apart and seas had flowed into the widening rift. They were trying to envision a mechanism whereby the seafloor could actually split open. This evidence finally appeared, but not until the early 1960s—three decades after Wegener's death.

Oceanographers measuring the paleomagnetic properties of rocks of the Atlantic Ocean floor were astonished to find a repeating series of rocks with alternating magnetic polarities: one stripe of rock with the same polarity (same magnetic north pole) as the Earth's present-day magnetic field, and the next stripe with the opposite polarity (north and south magnetic poles reversed). Scientists call these **normal** and **reversed magnetic polarities.** The stripes are hundreds of kilometers long, and they are exactly symmetrical on either side of the mid-ocean ridge that runs down the middle of the Atlantic Ocean. In other words, if you could fold the seafloor in half along the midocean ridge, the pattern of alternating paleomagnetic stripes on either side would match exactly.

What could this possibly mean? At first, scientists were mystified by these symmetrical patterns of magnetic stripes in seafloor rocks. Then two groups of geologists, working independently, came up with the same explanation. They proposed that the seafloor had split apart along the midocean ridge and that the rocks on either side were moving away from the ridge (Figure 1.5). As the rocks spread apart, lava from below welled up into the crack, solidifying into new volcanic rock on the seafloor. When the molten lava solidified, it took on the magnetic polarity of the Earth at that time. Over time, the spreading seafloor acted like a conveyor belt, carrying the newly magnetized rock away from the centerline of the ridge in either direction. This process came to be known as **seafloor spreading.** The discovery of seafloor spreading was probably the single most powerful piece of evidence in support of the hypothesis of continental drift.

Geologists have shown that the ages of seafloor rocks increase with distance from the midocean ridge. The youngest rocks are located along the centerline of the ridge, where new lava rises through the crack to the seafloor. The farther the rocks have moved from the ridge, the older they are. Every half-million years or so, for reasons that are not entirely understood, the Earth's magnetic field reverses itself—north becomes south, and south becomes north. (This is discussed in further detail in chapter 3.) As magnetic reversals occurred in the past, the changing polarities were recorded in newly forming rocks along the midocean ridge. The result was symmetrical stripes of rock with alternating polarities—normal,

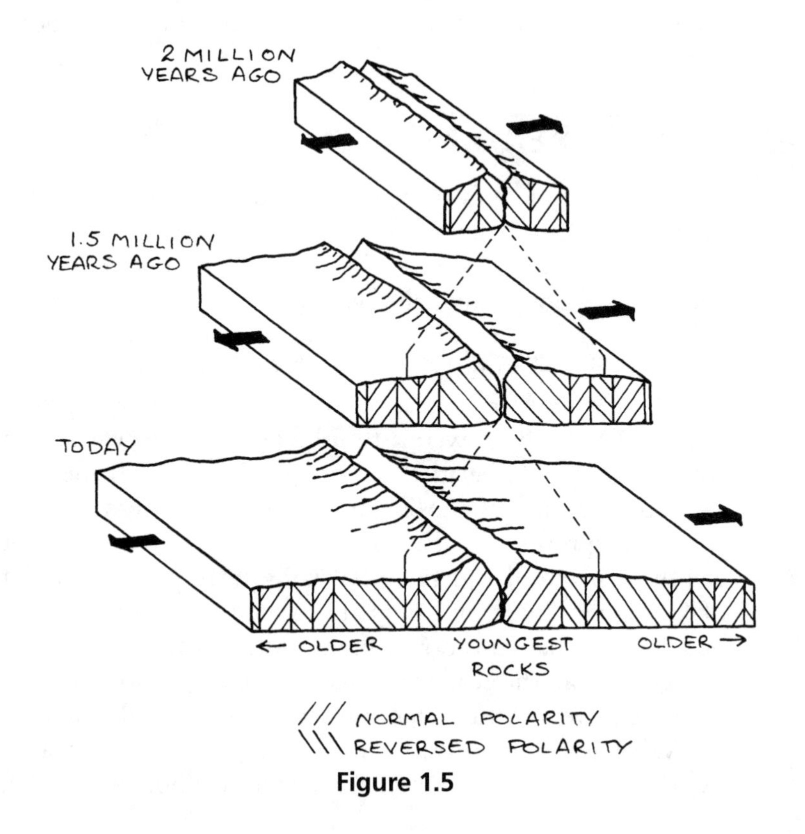

Figure 1.5

reversed, normal, reversed. This final piece of evidence convinced the great majority of geologists that seafloor spreading had indeed occurred and that the continents had drifted from their original locations.

What is the difference between a rock with normal paleomagnetic polarity and one with reversed paleomagnetic polarity?

___

___

___

Answer: A rock with normal polarity has a magnetic north pole in essentially the same orientation as the present-day North Pole. A rock with reversed polarity has magnetic north and south poles reversed from their present-day orientations.

10 PLATE TECTONICS IN A NUTSHELL

By the 1960s most scientists had become convinced that continental drift had really occurred. However, it remained to put all of this together into a coherent model. This model became the theory of plate tectonics. (It's called a theory now, instead of a hypothesis. A **hypothesis** is an educated guess; a **theory** is supported by extensive scientific evidence and testing.) Here is a brief summary of the theory of plate tectonics.

The outermost, rocky part of the Earth is the **crust.** As mentioned above, there are two types of crust: **continental crust,** which is relatively thick (average thickness 45 km, or 30 mi) and mostly made of granite, and **oceanic crust,** which is relatively thin (average thickness 8 km, or about 5 mi) and mostly made of basalt. Beneath the crust is the **mantle,** also made of rocks, but different from the rocks of the crust. At the center of the Earth is the **core,** made of iron-nickel metal, not rock. (You will learn more about the internal structure of the Earth in chapter 4.) Together, the crust and the outermost part of the mantle make up the **lithosphere,** a thin, cold, brittle, rocky layer (about 100 km, or 60 mi, thick, on average). The mantle below the lithosphere is very hot, so it is relatively malleable, like putty, even though it is made of solid rock. The part of the mantle immediately beneath the lithosphere is called the **asthenosphere;** it is especially weak because it is close to the temperature at which rocks begin to melt.

If you were to do an experiment in which you placed a very thin, cool, brittle shell (like the lithosphere) on top of hot, weak material that is rather squishy (like the asthenosphere), what would happen? You might predict that the thin shell would break into pieces. That is precisely the state of the Earth's lithosphere; it has broken into many large fragments, or plates. Today there are six large lithospheric plates, each extending for several thousands of kilometers, and a large number of smaller plates. The plates are in a condition called **isostasy,** which means that they are essentially "floating" on the weak asthenosphere, like blocks of wood floating on water.

You can experiment with plate motion by carefully heating wax in a pan and then letting it cool until it forms a thin skin or crust. Be careful—molten wax is very hot.

Think again about thin, brittle fragments floating on top of hot, squishy material. You might expect that movement in the underlying material would cause the brittle fragments to shift about. Again, that is exactly what happens to the Earth's lithospheric plates. As movement occurs in the hot mantle, the plates shift and interact with one another. Such movements involve complicated events that are collectively described by the term "tectonics" (from the Greek word *tekton,* meaning "carpenter" or "builder"). "Plate tectonics" thus refers to the study of the movement and interactions of lithospheric plates.

What is the lithosphere?

____

____

Answer: The outer 100 km (60 mi) of the Earth; the crust and the upper part of the mantle, above the asthenosphere.

11 PLATE MARGINS AND INTERACTIONS

Lithospheric plates interact with one another mainly along their edges, or margins. Plates can interact in three basic ways: they can move away from each other (diverge); they can move toward each other (converge); or they can slide past each other. Consequently, there are three kinds of plate margins: divergent, convergent, and transform fault margins.

Divergent margins are huge fractures in the lithosphere where plates move apart from one another (Figure 1.6A). When oceanic crust splits apart, seafloor spreading occurs and a **midocean ridge** is formed, like the one in the middle of the Atlantic Ocean. When continental crust splits apart, a great **rift valley** forms, as in East Africa where the African Plate is being stretched and torn apart. Eventually, a new ocean may form in the widening continental rift valley; a modern example of this is the Red Sea. In both continents and oceans, divergent margins are characterized by earthquakes (caused by the splitting and cracking of the rocks) and volcanism (caused by melted rock from the mantle welling up into the fractures).

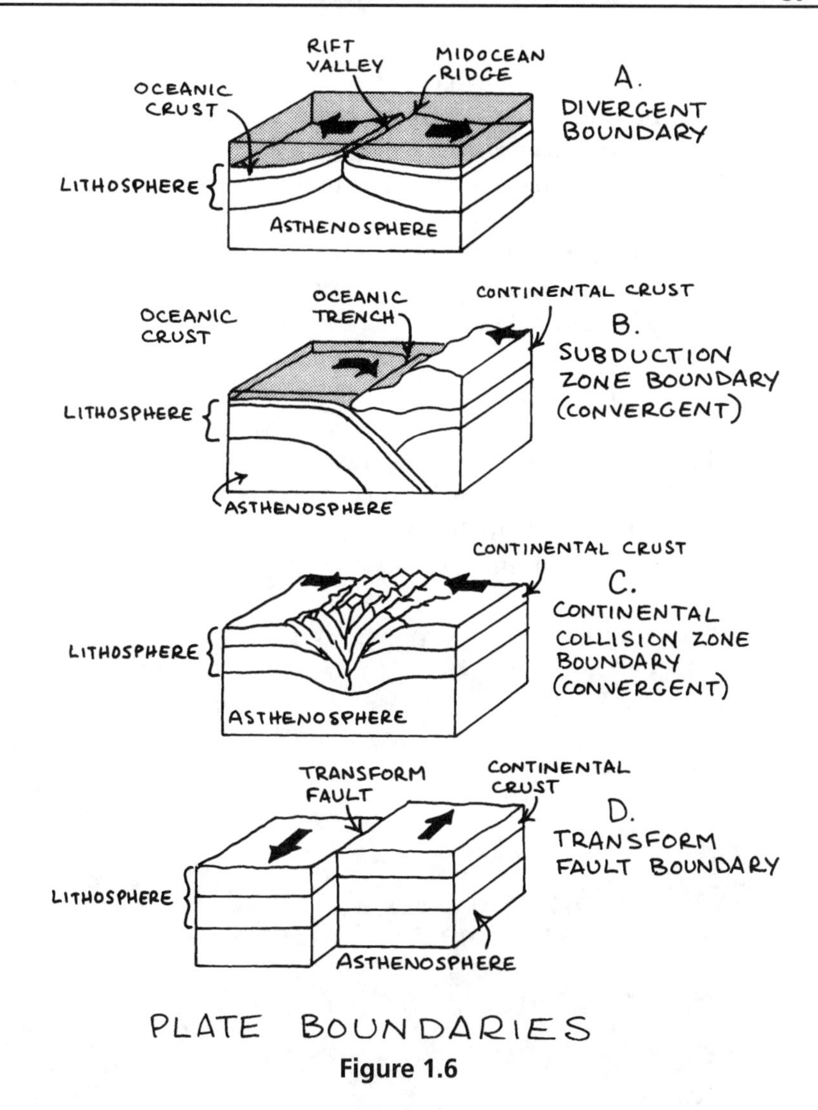

Figure 1.6

Convergent margins occur where two plates move toward each other. There are three basic types of convergent margins: ocean-ocean, ocean-continent, and continent-continent. Oceanic crust is made of basalt, which is denser (heavier) than the granitic rocks that make up the continental crust. Whenever oceanic crust is involved in a convergent margin, the dense oceanic crust sinks beneath the other plate (Figure 1.6B). This process is called **subduction,** and places where it occurs are called **subduction zones.** Subduction zones are marked by deep oceanic trenches and lines of volcanoes, as in Indonesia (an ocean-ocean subduction zone) or the Andes (an ocean-continent subduction zone).

When one continent meets another continent at a convergent margin, no oceanic crust is available to form a subduction zone. Instead, the continents collide and crumple up, forming huge, uplifted mountain ranges like the Himalayas; this is a **collision zone** (Figure 1.6C). Collision zones and subduction zones have lots of earthquake activity, caused by rocks colliding and grinding past one another.

Transform fault margins are huge fractures in the lithosphere where two plates slide past each other, grinding along their edges and causing earthquakes as they go (Figure 1.6D). A famous modern example is the San Andreas Fault in California, where the Pacific Plate is moving north-northwest relative to the North American Plate.

All these types of plate interactions are occurring today, as they have occurred throughout most of Earth's history. We don't often notice plate motion because lithospheric plates move very slowly—usually between 1 and 10 centimeters (cm) (0.4 to 4.0 inches [in]) per year. But we often feel the earthquakes and observe the volcanic activity that happens along active plate margins. The scars

CONTINENTAL RIFT. This is the Great Rift Valley, part of the East African Rift System in Kenya. The view is looking west across the Kerio River Valley from the top of the Tugen Hills to the Elgeyo Escarpment, which is about 1,220 meters (m) (4,000 feet [ft]) high. The Great Rift Valley was formed by tensional forces that stretched, thinned, and fractured the Earth's crust. (Courtesy Brian Skinner)

and remnants of ancient plate interactions are also preserved in the rock record for us to study.

What is the difference between a collision zone and a subduction zone?

____

____

____

Answer: A collision zone occurs where two continents converge. A subduction zone occurs along ocean-continent or ocean-ocean convergent margins.

12 CONVECTION: THE DRIVING FORCE

The theory of plate tectonics has been accepted by almost all geologists, but some questions remain. What causes plate motion? How does the mantle interact with the crust? What makes subduction occur? Scientists have a basic understanding of these processes, but the details have not been completely worked out. We know that thermal movement in the mantle is at least partly responsible for the movement of lithospheric plates. We also know that movement in the mantle is caused by the release of heat from inside the Earth. Let's examine the Earth's heat-releasing processes and consider how they cause plate motion.

The temperature inside the Earth is high—about 5,000° Celsius (C) (more than 9,000° Fahrenheit [F]) in the core. Some of this heat is left over from the Earth's beginnings, but some of it is constantly being generated by the decay of radioactive elements inside the Earth. This heat must be released; if it was not, the Earth would eventually become so hot that its entire interior would melt.

Some of the Earth's internal heat makes its way slowly to the surface through **conduction,** in which heat energy passes from one atom to the next. However, conduction is a slow way to transfer heat. It is faster and more efficient for a packet of hot material to be physically transported to the surface. This is similar to what happens when a fluid boils on a stovetop, as in the wax experiment described above. If you watch a fluid such as wax or spaghetti sauce as it boils, you will see that it turns over and over. Packets of hot material rise from the bottom of the pot to the top. As it reaches the surface, the hot fluid cools and then sinks back down to the bottom of the pot, where it is reheated. The continuous motion of material from bottom to top and down again is called a convection "cell," and this type of heat transfer is called **convection.**

Even though the Earth's mantle is mostly solid rock, it is so hot that it releases heat by convection (Figure 1.7). Rock deep in the mantle heats up and expands, making it buoyant. As a result, the rock moves toward the surface—very, very slowly—in huge convection cells of solid rock. Near the surface, the hot rock

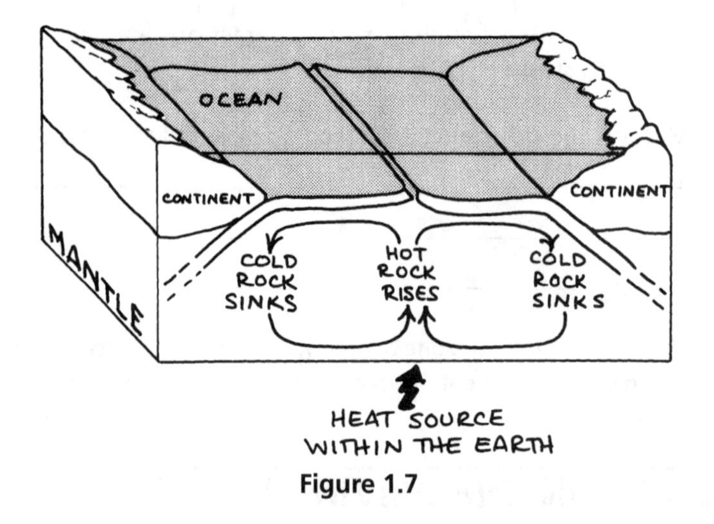

Figure 1.7

moves along the surface while losing heat, like the spaghetti sauce. The movement of hot rock in the asthenosphere is thought to be the main cause of plate motion. As the rock cools, it becomes denser (cool rock is denser, or heavier, than hot rock) and sinks back into the deeper parts of the mantle. This convection cycle provides an efficient way for the Earth to rid itself of some of its internal heat. Convection and the movement of plates near the surface create some of the most distinctive geologic and topographic features of the Earth's surface: the deep trenches where oceanic plates are subducted into the mantle; the midocean ridges and continental rift valleys where plates split apart; and the high, folded-and-crumpled mountain chains where continents collide.

Convection in the mantle is not nearly as simple as convection in a pot on a stovetop. Some of the most challenging unanswered questions about plate tectonics have to do with the exact nature of this process. Does the whole mantle convect as a unit, or is the top part of the mantle convecting separately from the bottom? In subduction zones, are the plates dragged down into the mantle, or do they sink under their own weight? What is the exact shape of convection cells in the mantle? Scientists are still seeking the answers to these and other questions about plate tectonics.

What is the temperature in the Earth's core?

Answer: About 5,000°C (9,000°F).

You have covered an enormous amount of material in this chapter—an entire scientific revolution in a few pages! Many of the concepts presented in this chapter may seem difficult and unfamiliar to you now, but don't worry. Plate tectonics

CONTINENTAL COLLISION. Mount Everest (center) is flanked by Lhotse (right) and Nuptse (left). They are the giants of the Himalaya Mountains, crowning the world's highest and most dramatic mountain range. The Himalayas were formed by compressive forces that folded, squashed, and thickened the Earth's crust. (Courtesy Brian Skinner)

is the foundation for our understanding of the Earth and its processes and materials, so many of these ideas will be revisited in the chapters to follow. Now test your knowledge of this material by trying out the Self-Test.

SELF-TEST

These questions are designed to help you assess how well you have learned the concepts presented in chapter 1. The answers are given at the end.

1. The ages of seafloor rocks generally ___________ with distance from a mid-ocean ridge, on either side of the ridge.
   a. increase
   b. decrease
   c. stay the same
   d. vary irregularly

2. The San Andreas Fault in California is a modern-day example of a ___________ plate margin.
   a. collisional
   b. subduction zone
   c. divergent
   d. transform fault

3. The "plates" in plate tectonics are made of fragments of ___________.
   a. continents
   b. oceanic crust
   c. the lithosphere
   d. the mantle

4. The weak layer of the mantle, immediately underlying the lithosphere, is called the ___________.

5. The Earth has two fundamentally different types of crust: ___________ crust is made mainly of basaltic rocks, and ___________ crust is made mainly of granitic rocks.

6. Along a noncliffed shoreline, the land usually slopes very gently toward the sea; this gently sloping land is called the abyssal plain. (T or F)

7. Wherever there is a convergent plate margin, a subduction zone will develop. (T or F)

8. Convection is faster and more efficient than conduction as a mechanism of heat transfer. (T or F)

9. Is the coastline, where the land meets the water, the true edge of a continent? Why, or why not?

____

____

____

10. What is the difference between a hypothesis and a theory?

____

____

____

11. Why is it tricky to match rock types and other geologic features across a continental split, such as where South America and Africa were once joined?

____

____

____

12. How does the distribution of glacial deposits support the idea that the continents were once joined together in the supercontinent Pangaea?

____

____

____

13. Summarize the main types of plate margins.

____

____

____

ANSWERS

1. a
2. d
3. c
4. asthenosphere
5. oceanic; continental
6. F
7. F
8. T
9. No. The true edge of a continent is where continental crust meets oceanic crust, but this is usually covered by mud (and sometimes by water). We define the edge of a continent to be halfway down the continental slope.
10. A theory is supported by extensive scientific evidence and testing; a hypothesis is an educated guess.
11. Rock-forming processes never cease. Some rocks were formed before the continents were joined, some while they were joined, others during the splitting of the continents, and still others after they became separated.
12. When the continents are rotated back into the "joined" position (Pangaea), the geology and ages of glacial deposits match remarkably well across the joins. Glacial grooves and scratches show that the ice was moving outward in all directions from what was then the South Pole.

13. The main types of plate margins are divergent (oceanic or continental); convergent subduction zone (ocean-ocean or ocean-continent); convergent collision zone (continent-continent); transform fault.

KEY WORDS

abyssal plain
apparent polar wandering
asthenosphere
collision zone
conduction
continental crust
continental drift
continental rise
continental shelf
continental slope
convection
convergent margin
core
crust
divergent margin
Earth system science
geology
historical geology
hypothesis
isostasy
lithosphere
mantle
midocean ridge
normal magnetic polarity
oceanic crust
paleomagnetism
Pangaea
physical geology
plates
plate tectonics
polarity
reversed magnetic polarity
rift valley
seafloor spreading
subduction
subduction zone
theory
transform fault margin

2 What the Earth Is Made Of

It isn't the size that counts so much as the way things are arranged.

—E. M. Forster

OBJECTIVES

In this chapter you will learn

- how our solar system formed and what kinds of objects make up our solar system;
- why the Earth is unique, as far as we know;
- how the basic building blocks of everything in and on the Earth—elements, atoms, and ions—combine to form chemical compounds;
- what types of minerals, rocks, and other materials make up the Earth.

1 OUR HOME PLANET

As geologists we mainly study our home planet, Earth. Sometimes it is helpful to broaden our perspective and take a look at Earth's place among its neighbors in space. Earth is one of nine planets in our **solar system**—the Sun and the group of objects orbiting around it. The solar system also includes more than 60 moons, a vast number of asteroids, millions of comets, and innumerable floating fragments of rock and dust. (When such fragments collide with a planet, we call them **mete-**

Meteorite, a rocky fragment from space. This specimen, which fell in 1960, near Bruderheim, Alberta, is about 3 inches across and composed mainly of two silicate minerals, olivine and pyroxene, plus a few grains of metallic iron. The specimen is covered by a dark fusion crust formed by the intense heat generated when the meteorite plunged through the atmosphere at supersonic speed. The true color of the meteorite is apparent on the cut made for scientific study of the specimen. It's composition is thought to be like that of the Earth's mantle, and the meteorite is thought to have come from a small planet between Mars and Jupiter that was later broken up by impacts and the gravitational force of Jupiter. Courtesy Brian Skinner.

orites.) All the objects in our solar system move through space in smooth, regular orbits, held in place by gravitational attraction. The planets, asteroids, and comets orbit the Sun, and the moons orbit the planets.

The planets can be separated into two groups on the basis of their characteristics and distances from the Sun. The innermost planets—Mercury, Venus, Earth, and Mars—are small, rocky, and relatively dense. They are similar in size and chemical composition and are called **terrestrial planets** because they resemble *Terra* ("Earth" in Latin). The outer planets (except Pluto) are much larger than the terrestrial planets, but much less dense. These **jovian planets**—Jupiter, Saturn, Uranus, and Neptune—have very thick atmospheres of hydrogen, helium, and other gases. Pluto—the smallest of the nine planets and the farthest from the Sun—doesn't really fit into either of these planetary groups. It is much smaller than the jovian planets, but much less dense than the terrestrial planets. In many respects, Pluto is more like a large comet than a planet.

The formation of the solar system most likely began with a swirling cloud of gas and dust that slowly flattened into a disk (Figure 2.1). The cloud is called the **solar nebula,** and this theory of how the solar system was born is called the **nebular theory.** Near the center of the rotating disk, pressure and temperature were extremely high. There, in the newly forming Sun, hydrogen atoms were subjected to such high pressures and temperatures that they began to undergo nuclear fusion, a process in which two light atoms combine to form a heavier atom, resulting in the release of energy. This process is still going on; it is the source of the Sun's radiant energy.

Eventually, the outer edges of the solar nebula cooled enough that solid particles began to condense, just as snowflakes condense from water vapor. Little by little, the innumerable dust-sized particles formed by condensation throughout the rotating cloud began to collide with one another. They stuck together, forming

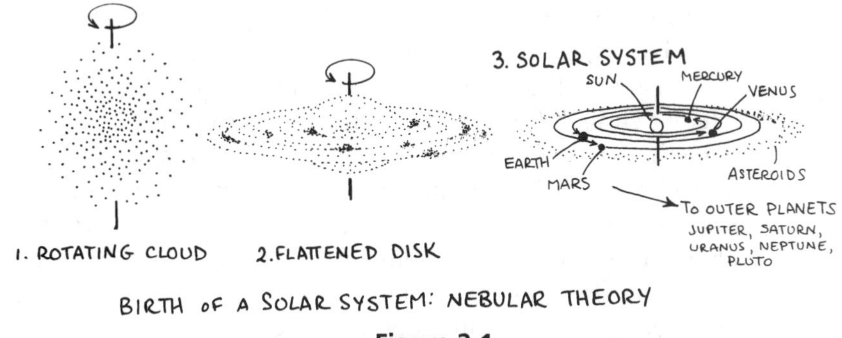

Figure 2.1

clusters. Through gravitational attraction, the largest clusters slowly swept up more and more particles. They grew into even larger clusters and eventually into the planets, moons, comets, and other solid objects of the solar system, including Earth. This growth process—the gradual gathering of solid matter into clusters—is called **planetary accretion.** The formation of Earth and other bodies in our solar system through condensation and accretion was essentially complete 4.56 billion years ago. (You can learn more about the solar system in *Astronomy: A Self-Teaching Guide,* by Dinah L. Moché.)

What are the main differences between the terrestrial planets and jovian planets?

____

____

____

Answer: The terrestrial planets are small and dense, with rocky compositions (similar to that of Earth). The jovian planets are larger and more massive, but much less dense than the terrestrial planets; they have thick hydrogen- and helium-dominated atmospheres.

2 WHAT MAKES EARTH UNIQUE?

The solar system is a group of planets and other objects that are related by the way they were formed and by their association with the Sun. Within this system is a smaller group, the terrestrial planets, which are related even more closely. The terrestrial planets have many things in common beyond their small sizes, rocky compositions, and positions close to the Sun. They have all been subjected to volcanism and intense meteorite impact cratering. They have all been hot and, indeed, partially molten at some time early in their histories. During this partially molten period, all of the terrestrial planets separated into three layers of differing chemical composition (as discussed in chapter 1): a relatively thin, low-density, rocky crust; a rocky mantle;

and a metallic, high-density core. This separation process is called **planetary differentiation;** it happened to all of the terrestrial planets, including Earth.

In spite of these similarities, the history and specific characteristics of Earth are different enough from those of the other terrestrial planets to make Earth habitable, while the others are not. If you look at a photograph of Earth taken from space, you immediately notice the blue-and-white **atmosphere,** an envelope of gases dominated by nitrogen, oxygen, argon, and water vapor, with traces of other gases. No other planet in the solar system has an atmosphere of this chemical composition.

Earth's atmosphere contains clouds of condensed water vapor. The clouds form because water evaporates from the hydrosphere, another unique feature. The **hydrosphere** ("watery sphere") consists of the oceans, lakes, and streams; underground water; and snow and ice. Planets farther from the Sun are too cold for liquid water to exist on their surfaces; planets closer to the Sun are so hot that any surface water evaporated long ago. Only Earth has just the right surface temperature to have liquid water, ice, and water vapor in its hydrosphere.

Another unique feature of Earth is the **biosphere,** the "life sphere." The biosphere comprises innumerable living things, large and small, which belong to millions of different species. It also includes recently dead plants and animals that have not yet completely decomposed.

The nature of Earth's solid surface is also special. Earth is covered by an irregular blanket of loose debris, which we call **regolith** (from the Greek *rhegos,* "blanket"). It forms as a result of the continuous chemical alteration and mechanical breakdown of rock through exposure to the atmosphere, hydrosphere, and biosphere. (You will learn more about this breakdown process, called weathering, in chapter 6.) Soils, muds in river valleys, sands in the desert, rock fragments, and all other unconsolidated debris are part of the regolith. Some other planets and planetary bodies with rocky surfaces are blanketed by loose, fragmented material, but in those cases the fragmentation has been caused primarily by the endless pounding of meteorite impacts. Earth's regolith is unique because it forms as a result of complex interactions of physical, chemical, and biological processes, usually involving water. It is also unique because it teems with life; most plants and animals live on or in the regolith or in the hydrosphere.

Below this blanket of loose debris is the solid rock of the lithosphere (the crust and the upper part of the mantle, as defined in chapter 1). Earth differs from all other known planets because of the unique relationship between its thin, brittle lithosphere and the hotter, weaker rocks that lie immediately below, in the asthenosphere. The solid rock that makes up the lithosphere is strong, but not strong enough to withstand the constant movement of underlying material caused by convection in the mantle. Consequently, plate tectonic activity has been an important process throughout much of Earth history. Plate tectonics is responsible for the uplifting of mountains, eruptions of volcanoes, intensities of earthquakes, and the shapes of continents and deep ocean basins. It has influenced the formation and chemistry of the atmosphere, the development of climatic zones, and the evolution of life.

So what makes Earth unique? We know of no other planet where plate tectonics has played, and continues to play, such an important role in forming the environment. We know of no other planet where water exists near the surface in solid, liquid, and gaseous forms. No other planet yet discovered would have been hospitable to the origin and evolution of life as we know it. There are billions upon billions of stars in the universe, so it is almost inevitable that there are billions of planets; surely a few of those planets must be earthlike and therefore capable of supporting life. However, if life does exist on a planet somewhere out in space, so far we haven't heard or seen any sign of it.

What is unique about the layer of loose, fragmented debris that blankets the surface of Earth?

____

____

____

Answer: Earth's regolith forms through a complex interaction of physical, chemical, and biological processes, usually involving water. Some other planets have regoliths, but on those planets the main weathering process is fragmentation by meteorite impacts.

3 BASIC BUILDING BLOCKS

Now let's have a closer look at the materials from which Earth (and all the other objects in our solar system) is built. You may have learned some fundamental concepts about chemical substances in high school chemistry, but it will be helpful to review them. (You can find a more comprehensive review in *Chemistry: Concepts and Problems, A Self-Teaching Guide,* by Clifford C. Houk and Richard Post.)

All matter on or in the Earth consists of one or more of the 92 naturally occurring chemical elements. An **element** is the most fundamental substance into which matter can be separated by chemical means. For example, table salt—the chemical compound sodium chloride, NaCl—is not an element because it can be separated into sodium (Na) and chlorine (Cl). But neither sodium nor chlorine can be further broken down chemically, so they are both elements. Every element is identified by a symbolic name. Some of these symbols, such as H for hydrogen, come from the element's name in English. Other symbols come from other languages. For example, the symbol for iron is Fe, from the Latin *ferrum;* the symbol for copper is Cu, from the Latin *cuprum,* which comes from the Greek *kyprios* (the island of Cyprus); and the symbol for sodium is Na, from the Latin *natrium.* The elements and their symbols are listed in Appendix 2.

A piece of a pure element, even a tiny piece no bigger than the head of a pin, consists of a vast number of identical particles called atoms. An **atom** is the smallest individual particle that has all the properties of a given chemical element. Atoms are so tiny that they can be seen only with the most powerful microscopes. Even then the image is imperfect because individual atoms are only about 10^{-10} meter (m) in diameter (that is, 0.00000000010 m).

Atoms are made of **protons,** which have positive electrical charges; **neutrons,** which are electrically neutral; and **electrons,** which have negative electrical charges (Figure 2.2). Protons and neutrons clump together to form the central **nucleus** of an atom, and electrons orbit the nucleus. The number of protons in the nucleus—its **atomic number**—is what gives an atom its special chemical characteristics. These characteristics identify it as a specific chemical element. Atomic numbers increase as we progress through the list of chemical elements, from hydrogen, atomic number 1 (one proton) to uranium, atomic number 92 (92 protons). The number of protons plus neutrons in the nucleus is called the **mass number** of the element.

The positive charge of a proton is exactly equal, but opposite, to the negative charge of an electron. Ideally, an atom has an equal number of protons and electrons and is therefore electrically neutral. An atom that has an excess positive or negative electrical charge caused by the loss, addition, or sharing of an electron is called an **ion.** When the charge is positive, the atom has given up electrons and we call it a **cation.** When the charge is negative, the atom has gained electrons and we call it an **anion.** A convenient way to indicate these charges is to record them as superscripts. For example, Li^+ (lithium) is a cation that has given up an electron, and so has a positive electrical charge; F^- (fluorine) is an anion that has gained an electron, and so has a negative electrical charge.

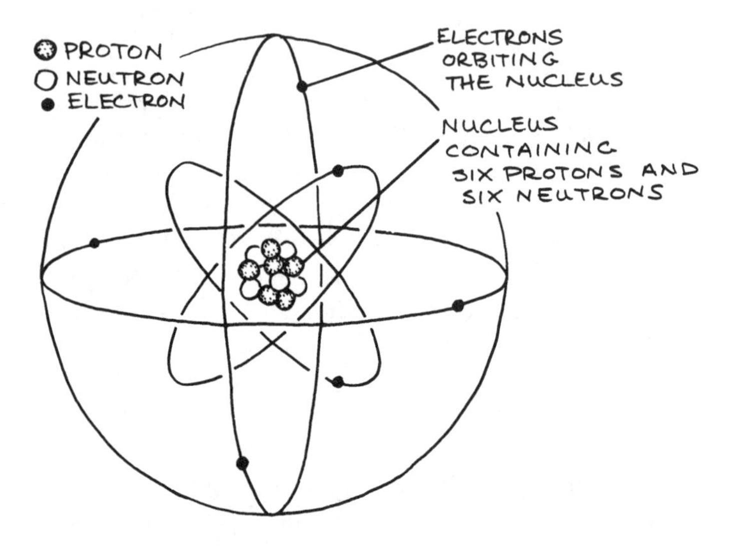

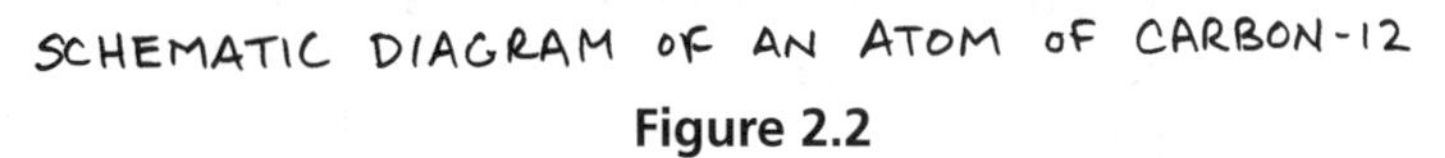

Figure 2.2

What is the atomic number of an element?

Answer: The atomic number is the number of protons in the element. It gives the element its specific identity.

4 BONDING AND COMPOUNDS

Chemical **compounds** form when one or more anions combine with one or more cations. For example, lithium and fluorine combine to form the compound lithium fluoride, LiF. We write it this way to show that for every Li^+ atom there is one F^- ion that exactly balances the electrical charge. Similarly, two cations of H^+ combine with one anion of O^{2-} to make the compound H_2O. The formula of a compound is written by putting the cations first and the anions second. The numbers of cations or anions in the compound are indicated by subscripts. For convenience, the charges (+ or −) are usually omitted. Thus, we write H_2O rather than $H_2^+O^{2-}$.

The properties of compounds are often quite different from those of their individual elements. For example, the elements sodium (Na) and chlorine (Cl) are highly toxic, but the compound NaCl (sodium chloride, the mineral halite, or table salt) is essential for human health. The smallest unit that has all the properties of a given compound is called a **molecule.** A molecule always consists of two or more atoms held together. The force that holds the atoms together is called **bonding.** There are several different types of chemical bonds. The type and strength of the bonds in a compound help determine the specific physical and chemical properties of that compound.

In each of the following compounds, which elements are cations and which are anions?

(a) MgO (b) FeS (c) CO_2 (d) $CaCl_2$ (e) K_2O

Answer:

(a) The cation is Mg (magnesium); the anion is O (oxygen).

(b) The cation is Fe (iron); the anion is S (sulfur).

(c) The cation is C (carbon); the anion is O (oxygen).

(d) The cation is Ca (calcium); the anion is Cl (chlorine).

(e) The cation is K (potassium); the anion is O (oxygen).

Why have you spent all this time learning about simple substances and compounds? Minerals—the building blocks of Earth materials—occur naturally as elements and chemical compounds. To understand minerals, you need to understand how they are put together from simple substances.

5 WHAT IS A MINERAL?

To be classified as a **mineral,** a substance must meet five requirements:

1. It must be naturally formed.
2. It must be solid.
3. It must be inorganic.
4. It must have a specific chemical composition.
5. It must have a characteristic crystal structure.

Let's consider these characteristics before we go on to discuss the specific properties of minerals.

The requirement that minerals be "naturally formed" excludes any substance produced in a laboratory, such as synthetic ruby, steel, glass, or plastic. Technically, none of these substances is a mineral. All liquids and gases—including naturally occurring liquids such as oil and natural gas—are excluded because minerals are solids. Ice in a glacier is a mineral but water in a stream is not, even though both are made of the same chemical compound, H_2O.

Minerals are "inorganic"; therefore, materials such as leaves and twigs, which come from living organisms and contain organic compounds, are not minerals. Coal is not a mineral because it is derived from the remains of plant material. The teeth and bones of dead animals and the shells of sea creatures, which are sometimes preserved as fossils, present a trickier case. Technically, teeth, bones, and shells are not minerals because they are formed by organic processes. However, when the remains become fossilized, the organic materials in them are replaced by minerals. If you were to perform a chemical analysis of a dinosaur bone, you would find only inorganic minerals, even though the bone's delicate internal structure may have been preserved almost intact.

Minerals have a "specific chemical composition." They occur either as simple chemical elements (gold, the element Au, is an example), or compounds with specific chemical formulas. Quartz, with the formula SiO_2, is an example of a compound with a specific chemical composition and formula. To a very limited extent, elements may be added to or substituted for the silicon and oxygen in the quartz, but if the chemical mixture strays too far from this formula, the mineral will no longer be quartz. It won't look like quartz, its internal arrangement of atoms will have changed, and it will no longer have the physical properties of the mineral we identify as quartz.

The fact that the chemical compositions of minerals have specific limits doesn't mean that their chemical formulas are all simple, like the formula for quartz. For example, the chemical formula of the mineral phlogopite is $KMg_3AlSi_3O_{10}(OH)_2$. Other minerals have even more complicated formulas. Nevertheless, the same thing is true for phlogopite as for quartz: if the chemical compound strays too far from its specific formula, the material will no longer have the same characteristics and will no longer be identifiable as phlogopite. The requirement of a specific chemical composition also serves to exclude materials whose composition cannot be expressed by an exact chemical formula. An example is glass, which is a mixture of many compounds and can have a very wide range of compositions.

Glass—even naturally formed volcanic glass—is further excluded from being a mineral by the requirement that minerals have "characteristic crystal structures." Technically, glass is a frozen liquid. The atoms in liquids (and in glass) are randomly jumbled, while the atoms in minerals are organized in regular, repeated geometric patterns (Figure 2.3). The geometric pattern of the atoms in a mineral is its **crystal structure.** The crystal structure of any mineral is a unique characteristic of that mineral; all specimens of that mineral have an identical crystal structure. Extremely powerful, ultrahigh-resolution microscopes enable scientists to look into the crystal structures of minerals and actually see the orderly arrangement of atoms in the mineral.

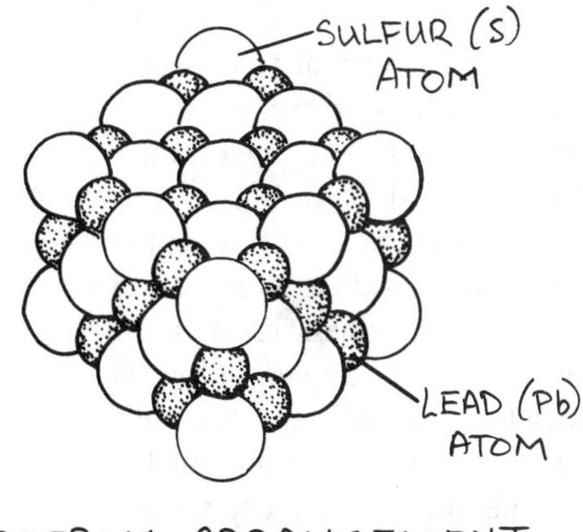

Figure 2.3

What are the five requirements that must be met for a substance to be a mineral?

___

___

Answer: The substance must be naturally formed, solid, and inorganic, and it must have a specific chemical composition and a characteristic crystal structure.

6 HOW TO TELL ONE MINERAL FROM ANOTHER

The properties of minerals are determined by their composition and crystal structure. Once we know the properties of minerals, we can use a few simple tests to identify them. It is usually not necessary to analyze a mineral chemically to discover its identity. The properties most often used to identify minerals are crystal form, habit, cleavage, hardness, luster, color, streak, and density. Let's look at each of these.

A **crystal** is any solid body that grows with flat ("planar") surfaces. The planar surfaces that bound a crystal are called **crystal faces,** and the geometric arrangement of crystal faces is called the **crystal form.** During the seventeenth century, scientists discovered that crystal form could be used to identify minerals, but they were unable to explain the wide variation in the sizes of crystal faces from one sample to another. Under some circumstances a mineral may grow into a long, thin crystal; under others, the same mineral may grow into a short, fat crystal. Why?

The person who solved this mystery was Nicolaus Steno. (Steno's real name was Niels Stensen. He often wrote in Latin, and his Latin name was Nicolaus Steno.) In 1669 he demonstrated that the unique property of a crystal of a given mineral is not the *size* of the faces, but rather the *angles* between the faces. The angle between any designated pair of crystal faces is constant, and it is the same for all specimens of a mineral, regardless of overall shape or size of the crystal or crystal faces (Figure 2.4). This is called Steno's law. Steno and others suspected that a mineral must have some kind of internal order that enables it to form crystals with constant interfacial angles. However, the particles on which that order depends—atoms—were too small for them to see. Proof that crystal form reflects internal order was finally achieved in 1912 when the German scientist Max von Laue used X rays to demonstrate that crystals are made of atoms packed in fixed geometric arrangements.

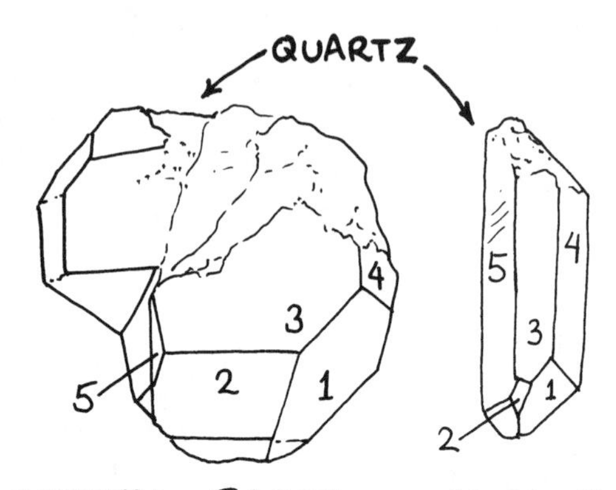

Figure 2.4

Crystal faces form mainly when minerals grow freely in an open space. Most minerals do not form in open, unobstructed spaces, so crystals with

CRYSTALS. These crystals of quartz (SiO_2) have grown unimpeded into an open space and so have developed crystal faces. The crystals are six-sided. The specimen is from Arkansas. (American Museum of Natural History)

well-developed faces are uncommon in nature. Instead, most minerals grow in limited spaces where other minerals get in the way. As a result, most mineral grains are irregularly shaped. However, in an irregularly shaped mineral grain—just as in a crystal—all the atoms are packed in the same strict internal geometric arrangement. In other words, crystals and irregular grains of a given mineral have identical crystal structures. That is why the term "crystal structure," rather than "crystal," is used in the definition of a mineral.

Some minerals grow such distinctively shaped grains that the shape of the grains—called the **habit**—can be used as an identification tool. For example, the mineral pyrite (FeS_2) commonly grows in the shape of a cube. The habit of chrysotile asbestos, a variety of the mineral serpentine, commonly takes the form of long, fine threads. Muscovite, a variety of mica, almost always grows in book-like stacks of clear, thin sheets.

The tendency of a mineral to break in preferred directions along planar surfaces is called **cleavage.** Most minerals break (or "cleave") more easily in some directions than in others. If you break a mineral with a hammer or drop it on the floor so that it shatters, some of the broken fragments will have surfaces that are smooth and planar, like crystal faces. But don't confuse crystal faces and cleavage surfaces, even though the two often look alike. A cleavage surface is a breakage surface, whereas a crystal face is a growth surface.

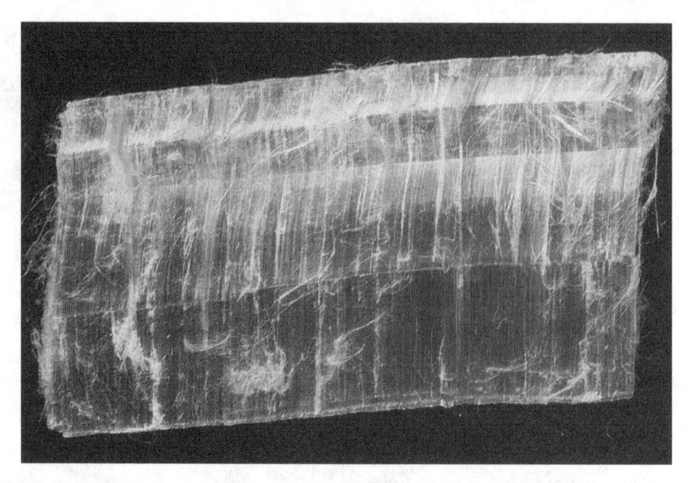

CRYSTAL HABIT. Some minerals have distinctive growth habits, even though they do not develop well-formed crystal faces. The mineral chrysotile sometimes grows as fine, cottonlike threads, as shown here, which can be separated and woven into fireproof fabric. When the mineral occurs like this, it is called asbestos. Chrysotile is one of several different minerals that are mined and commercially processed as asbestos. (Courtesy William Sacco)

MINERAL CLEAVAGE. The mineral halite (NaCl) has three well-defined cleavage directions, and it breaks into fragments bounded by three perpendicular faces. They look like crystal growth faces, but they are actually cleavage planes along which the mineral has broken. The cleavage pattern is controlled by the mineral's internal arrangement of atoms, its crystal structure. (Courtesy William Sacco)

The directions in which cleavage occurs are controlled by the in structure of the mineral. Cleavage takes place along planes where t between atoms are relatively weak. Because cleavage directions are directly to crystal structure, the angles between them are the same for all grains of a g mineral. Thus, just as the angles between the crystal faces are constant, so are th angles between cleavage planes. Crystals and crystal faces are rare; however, almost every mineral grain you see in a rock shows one or more breakage surfaces. That is why cleavage is a useful aid in the identification of minerals.

Hardness refers to a mineral's resistance to scratching. Hardness—like habit, crystal form, and cleavage—is governed by crystal structure and by the strength of the bonds between atoms. The stronger the bonds, the harder the mineral. Note the difference between hardness and cleavage; a mineral might be hard—that is, resistant to scratching—but still break or cleave easily.

Relative hardness values can be assigned by determining the ease or difficulty with which one mineral will scratch another. The **Mohs' relative hardness scale** has 10 steps, each identified by a common mineral. Talc, the basic ingredient of most body ("talcum") powders, is the softest mineral known; it is given a value of 1 on the scale. Diamond, the hardest mineral known, is given a value of 10, as shown in Table 2.1.

Any mineral on the scale will scratch all minerals below it. Minerals on the same step of the scale are just capable of scratching each other. For convenience, we often test relative hardness by using a common object such as a penny or a pocketknife as the scratching instrument, or glass as the object to be scratched.

Luster describes the quality and intensity of light reflected by a mineral. Luster may be "metallic," like that of a polished metal surface; "vitreous," like that of glass; "resinous," like that of resin; "pearly," like that of pearl; or "greasy," as if the surface was covered by a film of oil. Two minerals with almost identical color can have quite different lusters. Mineral colors are often striking, but color itself is not

Table 2.1 Mohs' Relative Hardness Scale

	Relative Hardness	Reference Mineral	Hardness of Common Objects
Softest	1	Talc	
	2	Gypsum	
	3	Calcite	Fingernail
	4	Fluorite	
	5	Apatite	Copper penny
	6	Potassium feldspar	Pocketknife, glass
	7	Quartz	
	8	Topaz	
	9	Corundum	
Hardest	10	Diamond	

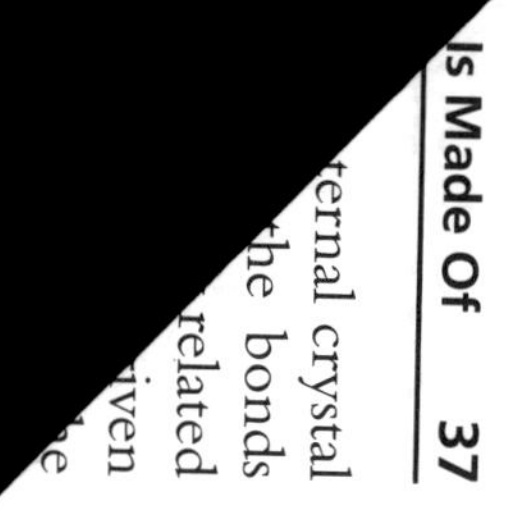

eans of identification. The color of a mineral is mainly determined by
›mposition. Some elements create strong color effects, even if they are
mpurities. For example, the mineral corundum (Al_2O_3) is commonly
ayish, but with a trace of chromium the corundum becomes blood red
n the name ruby. Traces of iron and titanium produce a deep-blue
called sapphire.
:an be particularly confusing in opaque minerals with metallic lusters.
use the color of such minerals is partly a property of the size of the mineral grains. One way to reduce variation is to prepare a sample of the mineral in which the grain size is uniformly fine. This is called a **streak,** a thin layer of powdered mineral made by rubbing the specimen across a nonglazed fragment of porcelain called a streak plate. The color of a streak is reliable because all of the grains in the streak are very small, so differences in color that might occur as a result of differences in grain size are minimized. For example, hematite (Fe_2O_3) always produces a red streak even though the specimen itself may look either black or red.

Another obvious physical property of a mineral is how light or heavy it feels. Two equal-sized baskets have different weights when one is filled with feathers and the other with rocks. The property that causes this difference is **density,** the mass per unit of volume of the material. Minerals with high density, such as gold, have closely packed atoms. Minerals with low density, such as ice, have loosely packed atoms. Density can be estimated by holding different minerals and comparing their weights. Metallic minerals generally feel heavy; most others feel light.

The densities of rocks and minerals are always reported in grams per cubic centimeter (g/cm^3). Notice that g/cm^3 contains a unit of mass or weight—the gram—per unit of volume—the cubic centimeter. A nonmetric equivalent unit would be pounds per cubic foot (lb/ft^3); however, this unit is not particularly familiar to most people in everyday usage. To get a better idea of what g/cm^3 means, you can compare the densities of minerals to that of water, which has a density of 1.0 g/cm^3 at the Earth's surface. In comparison, gold has a density of 19.3 g/cm^3 and feels very heavy. Many other minerals, such as galena (PbS) and magnetite (Fe_3O_4), which have densities of 7.5 and 5.2 g/cm^3, respectively, also feel heavy by comparison with other minerals. Most common minerals have densities in the range of 2.5 to 3.0 g/cm^3.

Why is color unreliable for mineral identification?

____

____

____

Answer: Color can vary widely in different samples of the same mineral, as a result of very small differences in chemical composition. Also, the grain sizes of minerals (especially opaque minerals) can affect the apparent color of a mineral sample.

7 THE ROCK-FORMING MINERALS

Geologists have identified approximately 3,500 minerals, but fewer than 30 of them are common in the crust of the Earth. (The most common minerals and their properties are listed in Appendix 3.) Why aren't there more minerals in the Earth's crust? The reason becomes clear when we consider the relative abundances of the chemical elements in the Earth's crust. Only 12 elements—oxygen, silicon, aluminum, iron, calcium, magnesium, sodium, potassium, titanium, hydrogen, manganese, and phosphorus—occur in the crust in amounts greater than 0.1 percent (by weight). Together, these 12 elements make up more than 99 percent of the mass of the crust (Figure 2.5). Therefore, it is no surprise that the crust is constructed of a limited number of minerals in which one or more of these 12 abundant elements is an essential ingredient. Minerals containing scarcer elements occur only in small amounts and only under special circumstances.

Two elements—oxygen and silicon—make up more than 70 percent of the crust by weight. Oxygen itself constitutes more than 60 percent of the crust in atomic proportion—that is, the actual number of atoms of oxygen in the crust—and more than 90 percent by volume. Oxygen is a large, lightweight atom; not only are there lots of oxygen atoms in the crust, they also take up a lot of space. Oxygen forms a simple anion, O^{2-}; compounds that contain this anion are called **oxides.** Oxygen and silicon together form an exceedingly strong anionic complex called a silica anion, $(SiO_4)^{4-}$; minerals that contain this anion are called **silicates.** Silicates are the most abundant of all minerals; oxides are the second most abundant. Other mineral groups based on different anions are less common.

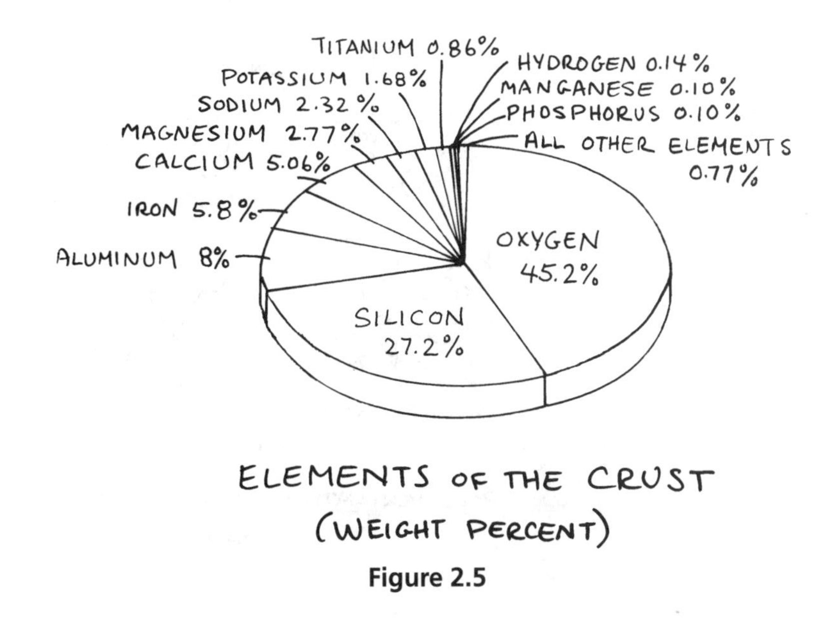

Figure 2.5

Silicate minerals, a few oxide minerals, and calcium carbonate ($CaCO_3$, the mineral calcite) make up the bulk of the Earth's crust—about 99 percent by volume. These common minerals are the **rock-forming minerals,** so called because they are the main components of all common rocks. Rock-forming minerals are everywhere—not only in rocks, but also in soils and sediment. Roads and buildings are constructed of rock-forming minerals; even the dust we breathe contains grains of these minerals. The two most common rock-forming minerals are quartz and feldspar; together, they constitute about 75 percent of the volume of the Earth's crust.

In silicate minerals, four oxygen atoms are tightly bonded to one silicon atom. The four large oxygen atoms make the corners of a tetrahedron, while the small silicon atom occupies the space at the center of the tetrahedron. Two silica tetrahedra can link together by sharing an oxygen atom at one of their points; this linking is called **polymerization.** Silicate tetrahedra can polymerize to form double tetrahedra, rings, chains, sheets, or three-dimensional frameworks (Figure 2.6). Different common cations (such as Ca^{2+}, Al^{3+}, Mg^{2+}, $Fe2^{2+}$, K^{+}, and Na^{+}) can fit into the spaces (interstices) between the linked silica tetrahedra. The identity and properties of a silicate mineral are determined by how the silica tetrahedra are linked together, which cations are present in the interstices, and how the cations are distributed throughout the crystal structure.

What are the two most abundant elements and the two most common rock-forming minerals in the crust of the Earth?

Answer: Oxygen and silicon; quartz and feldspar.

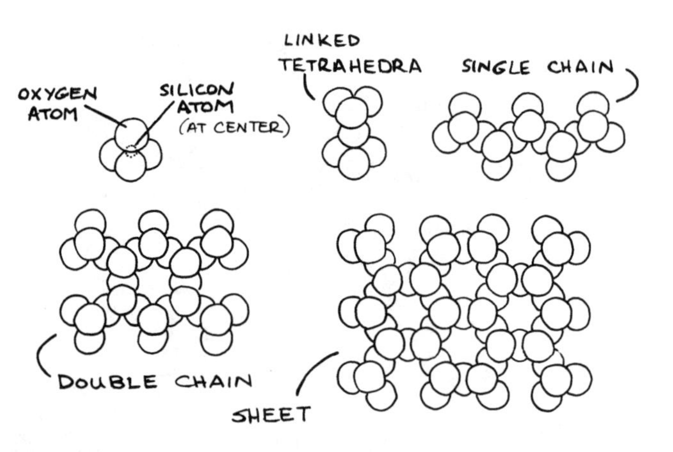

Figure 2.6

8 OTHER IMPORTANT MINERAL GROUPS

Oxides are the second most abundant group of minerals in the Earth's crust. Iron is one of the most abundant elements in the crust, and the iron oxides magnetite (Fe_3O_4) and hematite (Fe_2O_3) are the two most common oxide minerals. Magnetite takes its name from the ancient Greek word *magnetis,* meaning "stone of Magnesia," a town in Asia Minor. *Magnetis* had the power to attract iron particles, and it is the source of the word "magnet." The word "hematite" is derived from the red color of the mineral in powdered form; the Greek word for "red blood" is *haima.* Magnetite and hematite are the main minerals from which iron is derived for commercial use.

Another important mineral group is the **sulfides,** which contain the sulfur (S^{2-}) anion combined with different metal cations. Sulfide minerals typically are dense and have a metallic appearance. The most common are the iron sulfides pyrite (FeS_2, "fool's gold") and pyrrhotite (FeS). Many sulfides are ore minerals, which means that they are sought and processed for their valuable metal content. In chapter 11 you will learn about the geologic processes that concentrate minerals into ore deposits.

Like the silicates, oxides, and sulfides, each of the other mineral groups is based on combinations of cations with a particular anionic complex. The **carbonates** are based on the $(CO_3)^{2-}$ anionic complex; **phosphates** are based on $(PO_4)^{3-}$; and **sulfates** are based on $(SO_4)^{2-}$. Each of these groups includes important minerals. For example, apatite $[Ca_5(PO_4)_3(F,OH)]$*, a phosphate mineral, is the substance from which bones and teeth are made. Gypsum ($CaSO_4 \cdot 2H_2O$), a sulfate, is the raw material from which plaster is made. And the carbonate mineral calcite ($CaCO_3$), found in the shells of organisms such as mollusks, is the main constituent of limestone and marble.

A few materials occur in nature as **native elements;** that is, they are not combined with other elements. Minerals that occur in this form include some metals, such as gold (Au) and silver (Ag), and some nonmetals, such as sulfur (S), graphite (C), and diamond (C). Note that diamond and graphite have the same chemical composition—carbon (C)—but they have different crystal structures, so they are different minerals. Diamond and graphite are two **polymorphs** of carbon; the term "polymorph" refers to minerals that occur in many (*poly*) crystal forms (*morph*).

What anion or anionic complex forms the basis for each of the following mineral groups?

---

*The comma in (F,OH) means that fluorine (F^-) and hydroxyl (OH^-) can substitute freely for one another in the same location (called a "site") in the crystal structure.

(a) sulfides (b) carbonates (c) phosphates (d) sulfates (e) oxides (f) silicates

---

Answer: (a) S^{2-} (b) $(CO_3)^{2-}$ (c) $(PO_4)^{3-}$ (d) $(SO_4)^{2-}$ (e) O^{2-} (f) $(SiO_4)^{4-}$

9 WHAT IS A ROCK?

A **rock** is a naturally formed, coherent aggregate of minerals, which may be mixed with other solid materials such as natural glass or organic matter. The essential distinction between minerals and rocks is that rocks are *aggregates.* This means that rocks are collections of mineral grains (and sometimes other types of particles, such as organic material) stuck together. Rocks usually consist of several different types of minerals, but sometimes they are made of just one type of mineral. In either case, a rock will contain many grains of the constituent mineral or minerals. Sometimes the grains are held together by naturally formed cement. In other cases, the grains of the rock stick together because they have grown as crystals with interlocking grain boundaries.

Rocks are grouped into three families, which are defined and distinguished from one another by their properties and by the processes that form them. Here is a brief description of the three major rock families—igneous, metamorphic, and sedimentary rocks. You will learn more about each of them in subsequent chapters.

Igneous rocks (from the Latin *ignis,* "fire") are formed by the cooling and consolidation of magma, molten rock that is under the ground. If the magma reaches the surface while it is still in a molten state, it is called lava. As you will learn in chapter 5, some igneous rocks cool and crystallize slowly, deep under the ground; these are called **plutonic rocks.** Others reach the surface as lava and crystallize quickly; these are **volcanic rocks.**

The second major rock family is the **sedimentary rocks.** Unconsolidated rock and mineral particles that are transported by water, wind, or ice and then deposited are called **sediment.** A special kind of sediment is **soil,** which consists of loose particles that have been altered by biological processes to form a material that can support rooted plants. Through processes that you will learn about in chapter 7, sediment eventually becomes sedimentary rock, either by chemical precipitation from water at the Earth's surface or by the cementation of sediment.

The third important rock family is the **metamorphic rocks,** whose original form and mineralogy have been altered as a result of high temperature, high pressure, or both. The term "metamorphic" comes from the Greek *meta,* meaning "change," and *morphe,* meaning "form"; hence, change of form. You will learn more about metamorphic rocks and how they form in chapter 8.

THE THREE MAIN FAMILIES OF ROCKS. Granite (left) is an igneous rock formed by the cooling and crystallization of magma. Sandstone (right) is a sedimentary rock formed by the cementing together of grains of quartz. Gneiss (center) is a metamorphic rock; originally either a sedimentary or an igneous rock, the gneiss was deformed and given a layered fabric as a result of directed forces, high pressures, and elevated temperatures. (Courtesy Keith Stowe)

What is the difference between a mineral and a rock?

Answer: A rock is an aggregate (that is, a collection) of many mineral grains.

Once again, we have covered a lot of ground in this chapter. Now you know some things about how our solar system formed, about how our home planet differs from the other bodies in our solar system, and about the wide variety of materials that make up this planet. Try testing yourself on this information.

SELF-TEST

These questions are designed to help you assess how well you have learned the concepts presented in chapter 2. The answers are given at the end.

1. The theory that the solar system originated through condensation from a rotating cloud of gas and dust is called ___________.
   a. the nebular theory
   b. nuclear fusion

c. planetary differentiation
d. planetary accretion

2. Volcanic glass is not technically considered to be a mineral because it ___________.
   a. lacks a crystal structure
   b. is not naturally occurring
   c. does not have silicon or oxygen in its chemical composition
   d. All of the above are true

3. The Earth's crust is dominated by ___________ minerals.
   a. oxide
   b. carbonate
   c. silicate
   d. metallic

4. The binding force between elements in a chemical compound is called ___________.
   a. cement
   b. bonding
   c. crystal structure
   d. cleavage

5. The softest mineral on Mohs' relative hardness scale is ___________; the hardest is ___________.

6. The process whereby Earth (and some other planetary bodies) separated into concentric layers of differing composition is called ___________.

7. Nicolaus Steno discovered that different crystals of the same mineral will always have the same number of crystal faces, although the sizes of the faces and the angles between the faces may vary. (T or F)

8. Metamorphic rocks are those that have crystallized from a magma, deep within the Earth. (T or F)

9. C^{4+} combines with O^{2-} to form a common gas. Write the chemical formula for the gas. Which element is the cation and which is the anion in this chemical compound?

____

____

10. What are polymorphs?

____

____

11. There are many naturally occurring elements, and they form a wide variety of minerals (about 3,500 are known), yet there are only 30 common rock-forming minerals. Why is this so?

____

____

____

____

12. What is soil?

____

____

13. What is polymerization, and how does it occur in silicate minerals?

____

____

____

ANSWERS

1. a
2. a
3. c
4. b
5. talc; diamond
6. planetary differentiation
7. F
8. F
9. CO_2 (carbon dioxide); carbon is the cation, oxygen is the anion.
10. Polymorphs are minerals that have the same chemical composition but different crystal structures.
11. Only 12 elements (oxygen, silicon, aluminum, iron, calcium, magnesium, sodium, potassium, titanium, hydrogen, manganese, and phosphorus) occur in the crust in amounts greater than 0.1 percent (by weight). These 12 elements make up more than 99 percent of the mass of the crust. The crust is constructed of a limited number of rock-forming minerals (approximately 30) in which one or more of these 12 abundant elements is an essential ingredient. Minerals containing scarcer elements occur only in small amounts and only under special circumstances.
12. Soil is a special kind of sediment, which consists of loose particles that have been altered by biological processes to form a material that can support rooted plants.

13. Polymerization occurs when molecules link together. In silicate minerals, the $(SiO_4)^{4-}$ silica tetrahedra can link together by sharing an oxygen atom at one corner, forming double tetrahedra, rings, chains, sheets, or three-dimensional framework structures.

KEY WORDS

anion
atmosphere
atom
atomic number
biosphere
bonding
carbonate
cation
cleavage
compound
crystal
crystal face
crystal form
crystal structure
density
electron
element
habit
hardness (of a mineral)
hydrosphere
igneous rock
ion
jovian planet
luster
mass number
metamorphic rock
meteorite
mineral
Mohs' relative hardness scale
molecule
native element
nebular theory
neutron
nucleus
oxide
phosphate
planetary accretion
planetary differentiation
plutonic rock
polymerization
polymorph
proton
regolith
rock
rock-forming mineral
sediment
sedimentary rock
silicate
soil
solar nebula
solar system
streak
sulfate
sulfide
terrestrial planet
volcanic rock

3 The Rock Record and Geologic Time

The mind seemed to grow giddy by looking into the abyss of time.

—John Playfair

OBJECTIVES

In this chapter you will learn

- how scientists know the age of the Earth;
- how the fundamental principles of stratigraphy and fossil correlation help us understand geologic time and ancient processes;
- how geologists used these principles to develop the geologic time scale;
- how scientists determine the numerical ages of rocks, fossils, and geologic events.

1 UNDERSTANDING GEOLOGIC TIME

In chapter 2 you learned that the age of the solar system, including Earth, is about 4.56 billion years. How do scientists know this? In this chapter you will find out how scientists measure the ages of rocks, geologic events, and Earth itself.

Try this thought experiment. Think of the whole of Earth history as though it were compressed into a single year. Earth and the solar system formed at midnight on New Year's Eve. On this scale, the oldest known

> terrestrial rocks would date from late February, and the oldest known fossils from late March. The first land plants and animals emerged near the end of November, and the first flowers appeared around November 20. Dinosaurs ruled the Earth in mid-December, but they were gone by December 26. The oldest known hominid (humanlike) fossils would date from shortly after noon on December 31, but our own species—*Homo sapiens*—didn't appear until about three minutes before midnight. Great thicknesses of ice covered much of Canada and the northern United States until just over a minute before midnight. The whole history of human civilization happened within the last 45 seconds of the year.

We use the year as the unit of time for both historic and geologic events, but many geologic events happened so long ago that the numbers of years seem astronomically large. To a geologist, an event that happened 1 million or even 10 million years ago is "recent." Appreciating the immensity of geologic time is a real challenge.

The first attempts to measure geologic time were made about two centuries ago. Geologists speculated that if they measured the amount of sediment transported by streams in a given period of time, it might then be possible to estimate the time needed to erode away an entire landscape. These attempts were imprecise, but they demonstrated that the Earth must be immensely ancient compared to human history. One of the founders of modern geology, James Hutton, wrote in 1785 that for the Earth there is "no vestige of a beginning, no prospect of an end." Hutton thought the Earth must have been here since the beginning of time and must surely go on forever.

The idea that present-day geologic processes, such as the transport of sediment by streams, can provide scientific insights into ancient geologic processes is a fundamental principle of geology. This principle is called **uniformitarianism,** and it was first stated by Hutton. The principle of uniformitarianism says that the processes operating in Earth systems today have operated in a similar manner throughout much of geologic time; it is another way of saying that "the present is the key to the past." We can examine any rock, however old, and compare its characteristics with those of similar rocks that are forming today. We can then infer that the ancient rock very likely formed in a similar environment, through similar processes, and on a similar time scale.

Geologists have used the principle of uniformitarianism to explain the Earth's geologic features and processes. In so doing, they have discovered that the Earth is incredibly old. An enormously long time is needed to erode a mountain range, or for huge quantities of sand and mud to be transported by streams, deposited in the ocean, and cemented into rocks, or for rocks to be uplifted to form a mountain. Yet the cycle of erosion, formation of new rock, uplift, and more erosion has been repeated many times during the Earth's long history.

What is the principle of uniformitarianism?

____

____

Answer: The geologic processes that we see today have been operating in much the same manner throughout most of Earth history; "the present is the key to the past."

2 RELATIVE AGE AND THE PRINCIPLES OF STRATIGRAPHY

The geologists who followed Hutton realized that the Earth is very ancient, but they lacked a way to determine exactly how old it is. All they could do was determine the **relative ages** of past events—that is, how old one rock formation or geologic feature was in comparison to another. In other words, they were concerned with establishing the chronologic sequence of events in Earth history. Nineteenth-century geologists built a geologic time scale based on relative ages. In doing so, they took the essential first steps toward unraveling the history of Earth.

The concept of relative age is based on a few simple but profound principles, which come from the study of sedimentary rock layers or **strata** (the singular form is **stratum,** from the Latin word for "layer"). The scientific study of rock strata is called **stratigraphy,** and it was through the principles of stratigraphy that the geologic time scale was developed.

The first fundamental principle of stratigraphy is the **principle of original horizontality,** which says that waterborne sediments settle out and are deposited (laid down) as horizontal layers. The principle of original horizontality is important because it means that whenever we observe water-laid strata that are bent, twisted, or no longer horizontal, we know that some tectonic force must have disturbed the strata after they were deposited.

Test the principle of original horizontality. Take a tub of water, stir in some mud, and let it settle quietly. Examine the result: the layers of deposited sediment will be horizontal, unless the tub was disturbed during the settling.

The second principle of stratigraphy is the **principle of stratigraphic superposition,** which states that any sedimentary rock layer is younger than the stratum below and older than the stratum above. In other words, sedimentary rock strata are like newspapers laid in a pile day by day, providing a record from the time of deposition of the bottom (oldest) stratum to the time of deposition of the top (youngest) stratum. Newspapers, of course, have dates on them, so it is possible to determine the exact age of a given newspaper within the pile. Rock strata lack dates, so on the basis of stratigraphy we can only assign relative ages to them.

Let's say you have been piling up your newspapers day by day, but on Tuesday you take the paper with you to read on the subway. When you have finished reading the paper, you throw it in the recycling bin at the subway station instead of carrying it home. This means that there will be a gap in the chronologic sequence within your pile of newspapers at home—Wednesday's paper will fall directly on top of Monday's paper, with a gap where Tuesday's paper would have been.

Similarly, there can be time gaps in sequences of rock strata. Imagine that a stream is transporting sediment and depositing it on the floor of a valley, one layer on top of another. Then imagine that the stream stops depositing sediment—perhaps it dries up or is diverted. The pile of sediments sits exposed in the valley for thousands or even millions of years. Some strata on top of the pile are worn away by erosion. Then imagine that a stream begins to deposit sediment on the pile of strata once again. If you were to examine the whole sequence of strata, you would

UNCONFORMITY. The surface of an unconformity between horizontal sedimentary rocks (above) and eroded granite (below) is clearly visible in this photo. After the granite cooled and crystallized, it was uplifted and exposed by erosion. The old surface of erosion is so nearly horizontal that it is followed by the modern roadway seen in the photo. Following erosion, a sequence of sediments was deposited unconformably on the old granite. The location is near Cape Town, Republic of South Africa. (Courtesy Brian Skinner)

find a gap representing the period during which the strata were exposed to erosion. Such a gap is called an **unconformity.** Sometimes it is difficult to spot an unconformity in a sequence of horizontal sedimentary rock strata, but unconformities are important because they can represent very long periods of time that are missing from the sedimentary rock record (Figure 3.1).

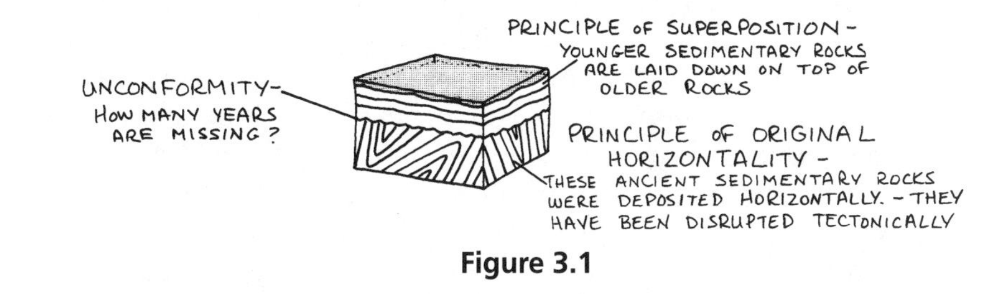

Figure 3.1

Another important concept in the determination of relative ages is the **principle of crosscutting relationships,** which states that a rock is always older than any feature that cuts or disrupts it. If a rock is cut by a fracture, for example, then the rock itself is older than the fracture that cuts it (Figure 3.2). Sometimes igneous rocks form when magma intrudes and cuts across previously existing sedimentary rock strata; in this case, too, the sedimentary rocks must be older than the igneous rocks that cut and disrupt the strata. The principle of crosscutting relationships is very helpful in determining the chronologic sequence of events.

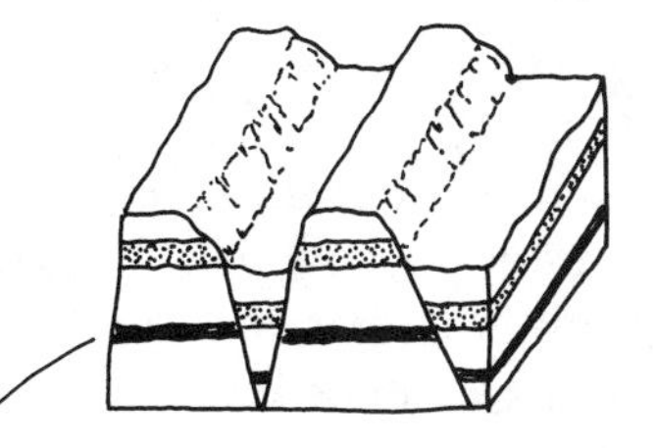

Figure 3.2

What is an unconformity?

____

____

____

Answer: A gap in a stratigraphic sequence, marking the absence of part of the geologic record. It typically represents a period of time during which there was no deposition and possibly some erosion of preexisting rocks.

3 FOSSILS AND STRATIGRAPHIC CORRELATION

Many sedimentary strata contain **fossils,** the remains of plants and animals that died and were incorporated and preserved as the sediment accumulated. Fossils can be shells, bones, leaves, twigs, or even the tracks and footprints of animals—anything that records the former presence of life. One of the first people to investigate fossils in a serious way was Niels Stensen, or Nicolaus Steno, the same person you learned about in chapter 2, who first determined that the angles between faces in crystals of the same mineral are constant. Steno was interested in the origin of fossil shark's teeth. He did not realize they were fossils when he started his study. Steno published his ideas in a paper in 1669, in which he articulated the principles of superposition and original horizontality. He stated further that fossils were the remains of ancient life. His conclusions were ridiculed at the time. But a century later, by Hutton's time, the fact that fossils are the remains of ancient plants and animals was widely accepted. The study of fossils and the record of ancient life on Earth is known as **paleontology.**

In Hutton's time there was a young surveyor named William Smith who was laying out the routes for canals in southern England. As Smith worked, he recorded the kinds of rock that were excavated and took note of the fossils in the various strata. In doing so, he made an important discovery: each group of strata contained specific assemblages of fossils. This was of practical significance because it meant that whenever Smith came across a new outcropping of rock, he could look at the fossils and immediately say where the stratum belonged in the chronologic sequence of rock strata. Smith used this approach to correlate strata from one

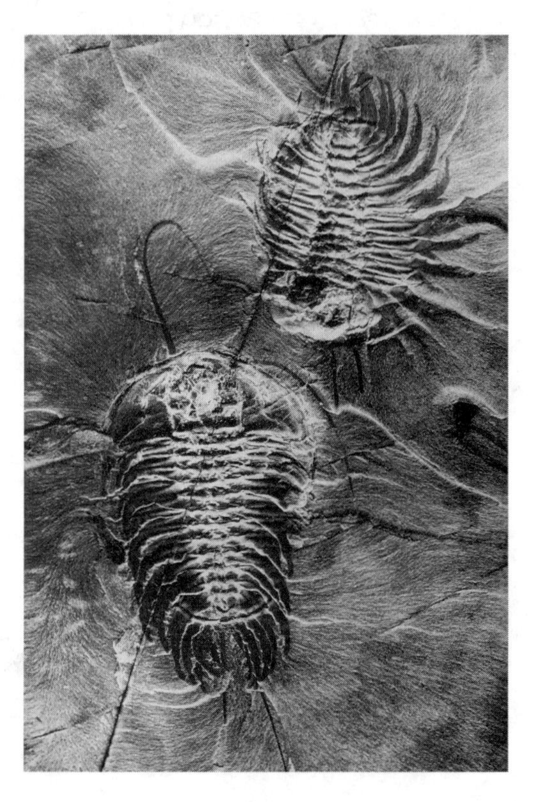

FOSSILS. Trilobites are ancient arthropods that once crawled on the seafloor. Long extinct, these fossil specimens are from the Cambrian period, more than 500 million years ago. (Smithsonian Institution)

place to another. Eventually, he could look at a specimen from any sedimentary rock in southern England and name the stratum from which it had come and the position of the stratum in the sequence. William Smith's practical discovery turned out to be of great scientific importance as well. What he had discovered was a means of **stratigraphic correlation**—that is, a method of equating relative ages in successions of strata from two or more different places. Initially, Smith correlated strata over a few kilometers; then he showed that it worked over distances of tens of kilometers. Geologists quickly recognized the scientific importance of Smith's work and soon carried out correlations between sequences of strata hundreds and even thousands of kilometers apart.

What is a fossil?

Answer: A fossil is the remains of a plant or animal that died and was incorporated and preserved in sediment as it accumulated.

4 THE GEOLOGIC COLUMN

One of the greatest successes of nineteenth-century science was the demonstration, through stratigraphic correlation, that sequences of rock strata are the same on all continents. This meant that a gap in the stratigraphic record in one place could be filled by evidence from somewhere else. Through worldwide correlation, nineteenth-century geologists assembled the **geologic column,** the succession of all known strata fitted together in chronologic order on the basis of their fossils or other evidence of relative age. Standard names are now used worldwide for the subdivisions of the geologic column (Figure 3.3). The geologic column is divided into four major time divisions called **eons.** The eons, which all have Greek names, are the **Hadean** ("beneath the Earth"), **Archean** ("ancient"), **Proterozoic** ("early life"), and **Phanerozoic** ("visible life"). The eons encompass hundreds of millions to billions of years. They are divided into shorter spans of time called **eras.** Eras are most useful in dividing up the Phanerozoic eon because they are defined by fossil assemblages; fossils are absent or very rare in rocks of the earlier eons. The three eras of the Phanerozoic eon are the **Paleozoic era** ("ancient life"), **Mesozoic era** ("middle life"), and **Cenozoic era** ("recent life").

During the Paleozoic era, life-forms on Earth progressed from marine invertebrates (animals without backbones) to fishes, amphibians, and reptiles. Early land plants appeared during the Paleozoic era. The Mesozoic era witnessed the appearance of the first flowering plants and the rise of dinosaurs. Dinosaurs were the dominant land vertebrates (animals with backbones) for millions of years. Mam-

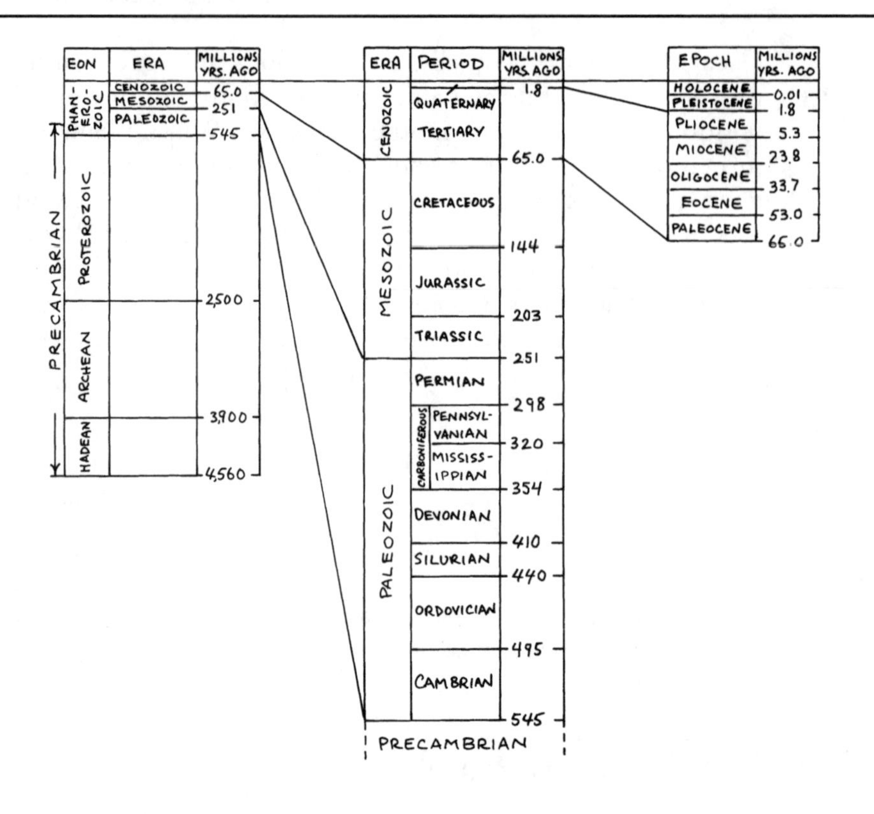

THE GEOLOGIC COLUMN

Figure 3.3

mals first appeared during the Mesozoic era, but they did not become dominant until the disappearance of the dinosaurs at the end of the Mesozoic era. During the Cenozoic era, grasses appeared and became important food for grazing mammals, the dominant land vertebrates.

The three eras of the Phanerozoic eon are further subdivided into shorter time units called **periods.** The geologic periods have been defined as a result of nearly a century of detailed work on fossil assemblages in the strata of Europe and North America. The earliest period of the Paleozoic era, the Cambrian period, is the time when animals with hard shells first appear in the geologic record. Before the Cambrian period all animals were soft-bodied or microscopic, and they didn't leave much fossil evidence. Geologists often refer to the entire time span and all rocks formed during the time preceding the Cambrian as simply **Precambrian.**

Periods lasted tens of millions of years, so they have been split into still smaller time divisions called **epochs.** The names of the epochs are mostly of importance to specialists. Some of the recent epochs are more familiar than others—such as the Pleistocene epoch, for example—because these are the times when humans and their ancestors emerged, and the names sometimes appear in the popular press.

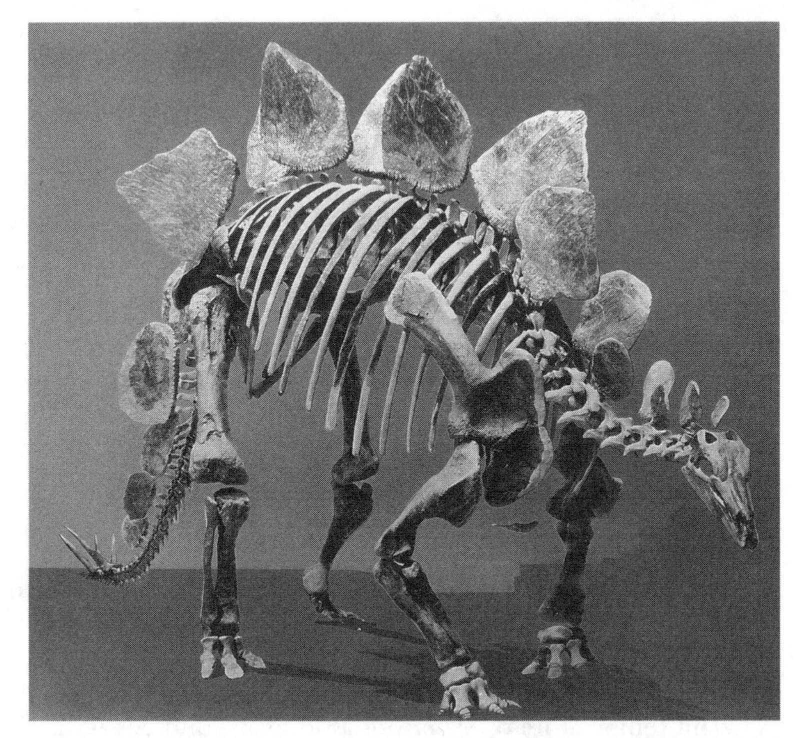

DINOSAUR. *Stegosaurus stenops* lived during the Jurassic period, about 150 million years ago, when dinosaurs ruled the Earth. This fossilized dinosaur skeleton, which stands almost 2 m (about 6 ft) in height, was found in the Morrison Formation, Albany County, Wyoming. (Smithsonian Institution)

What are the names of the four eons of the geologic column? List them in order from the oldest to the most recent.

Answer: Hadean, Archean, Proterozoic, and Phanerozoic.

5 NUMERICAL AGE

Scientists who worked out the geologic column were tantalized by the challenge of measuring **numerical age**—that is, the actual age of rocks and fossils in years (formerly called "absolute" age). They wanted to find specific answers to questions such as: How old is the Earth? How fast do mountain ranges rise? How long did the Paleozoic era last? How long have humans inhabited the Earth? To get answers, they needed a way to determine numerical age.

The need to measure time quantitatively became intense following the publication, in 1859, of Charles Darwin's controversial book, *On the Origin of Species by Means of Natural Selection*. Darwin understood that the evolution of new species must be a very slow process, requiring vast amounts of time. Opposing Darwin on the time issue was Lord Kelvin, a leading physicist of his time. Kelvin attempted to determine the numerical age of the Earth by calculating the amount of time it has been a solid body. The Earth started as a very hot object, he argued. Once it had cooled sufficiently to form a solid outer crust, it could continue to cool only through the loss of heat by conduction through solid rock. By measuring the thermal properties of rock and estimating the present temperature of the Earth's interior, he calculated the time for the Earth to cool to its present state. The answer he obtained for the age of the Earth was somewhere between 20 and 100 million years. Darwin realized that this provided too little time for natural selection to explain the fossil record as he believed it had happened, so in later editions of his book, he paid little attention to the time problem.

Darwin died before it was discovered that Lord Kelvin was wrong. Kelvin had assumed that no heat had been added since the Earth was formed. When he made these calculations, however, radioactivity was not known; now we know that natural radioactivity continuously supplies heat to the Earth's interior. Instead of cooling rapidly, the Earth's interior is cooling so slowly that it has a nearly constant temperature over periods as long as hundreds of millions of years.

What was needed to resolve the numerical time problem was a way to measure events by some process that runs continuously, that is not influenced by chemical reactions or high temperatures, and that leaves a continuous record without any gaps. In 1896, the discovery of radioactivity not only proved that Kelvin was wrong, but also provided the breakthrough for a technique to measure numerical time.

What is the difference between relative age and numerical (or "absolute") age?

____

____

____

Answer: Relative age is the age of a rock formation or geologic event in comparison to something else; that is, the rock or event is known to be older or younger than some other feature. Numerical age is the actual age (in years) of a rock or geologic event.

6 RADIOACTIVITY AND RADIOACTIVE DECAY

You learned in chapter 2 that the atomic number of an element—that is, the number of protons in the nucleus of the element—is constant and characteristic of that element. Most chemical elements exist naturally in two or more different forms, called **isotopes.** Each isotope of a given element has the same number of protons but a different number of neutrons in the nucleus. In other words, each isotope of an element has the same atomic number but a different mass number.

Most isotopes are stable, but a few, such as carbon-14 and uranium-238, are naturally unstable and their nuclei are subject to change. An isotope with an unstable nucleus that spontaneously changes its atomic number, its mass number, or both is said to be "radioactive." The process of change is referred to as **radioactive decay,** and the phenomenon is called **radioactivity.** An isotope undergoing radioactive decay is a "parent," and an isotope that is produced by radioactive decay is a "daughter." For example, carbon-14 (parent) decays to nitrogen-14 (daughter); uranium-238 (parent) decays to lead-206 (daughter). Radioactive decay involves the release of different kinds of radioactive particles—alpha (α) and beta (β) particles—and gamma (γ) rays. Radioactive decay also releases heat. It was the heat from the decay of radioactive isotopes inside the Earth that made Lord Kelvin's estimates of the numerical age of the Earth wrong.

The science of radioactivity is very complicated. But what is most important for our discussion is the fact that each radioactive isotope has its own measurable decay rate. For this reason, the radioactive decay of natural isotopes can serve as a naturally occurring clock, one that is built into rocks. Furthermore, a rock that contains several different radioactive isotopes has several built-in clocks that can be checked against one another. Careful study of radioactive isotopes in the laboratory has shown that decay rates are unaffected by changes in the chemical and physical environment. This is important because it means that the rate of radioactive decay of a given isotope is not altered by geologic processes like erosion or metamorphism.

In radioactive decay, the number of parent atoms continuously decreases as the number of daughter atoms continuously increases (Figure 3.4). The rate of decay is measured by the **half-life,** the time it takes for the number of parent atoms to be reduced by one-half. Let's say the half-life of a particular radioactive isotope is 1 hour. If we started an experiment with 1,000 parent atoms, after an hour 500 parent atoms would remain and 500 daughter atoms would have been formed. At the end of the second hour, another half of the parent atoms would be gone, so there would be 250 parent and 750 daughter atoms. After the third hour, another half of the parent atoms would have decayed, and so on. At any given time, the number of remaining parent atoms plus the number of daughter atoms equals the number of parent atoms that the rock or mineral originally started with. The proportion of parent atoms decaying during each time interval (each half-life) is 50 percent. The same kind of exponential law governs compound interest paid on a bank account.

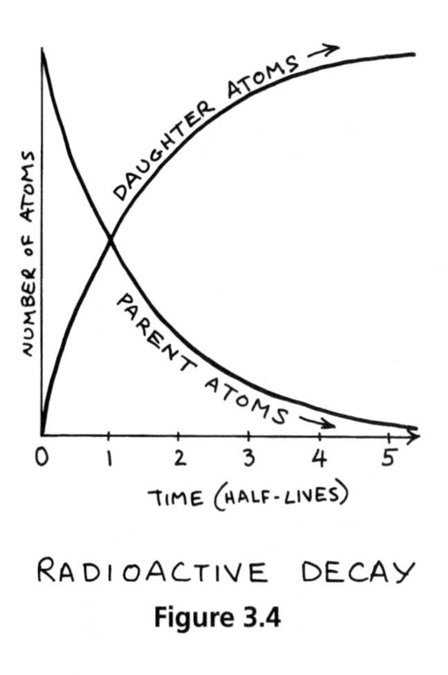

Figure 3.4

What is a half-life?

Answer: The time it takes for the number of parent atoms in a radioactive material to be reduced by one-half. (Note that the concept of half-life is also used in reference to materials that decay by processes other than radioactivity, such as biodegradation.)

7 RADIOMETRIC DATING

The steady, measurable rate of decay of radioactive isotopes is the key to their use in determining the numerical ages of rocks. The first determinations of the numerical ages of rocks using radioactivity were made in 1905. The long-hoped-for "rock clock" was finally available, and the results have been—and continue to be—remarkable. Our ability to measure numerical age has revolutionized the way we think about the Earth. The study of time in relation to the history of the Earth, and the determination of numerical ages using radioactivity, is called **geochronology.**

A radioactive rock clock works in the following way. The instant a new mineral grain forms, all of the atoms are locked in the crystal structure and removed from reaction with the environment outside the grain. The atoms in the mineral grain are, in a sense, sealed in an atomic bottle. If some of the atoms are radioactive, we can determine how long ago the bottle was sealed—that is, how long ago the mineral grain formed—by measuring the number of parent and daughter atoms that remain. The daughter atoms can only have come from radioactive decay of parent atoms. If we know the rate of decay (in terms of the half-life) of the radioactive parent, we can calculate how long ago the mineral grain formed. **Radiometric dating** is the use of naturally occurring radioactive isotopes to determine the time of formation of minerals and rocks. Radiometric dating is most useful for determining the time that minerals formed in igneous and metamorphic rocks.

The radiometric age of a mineral grain contained in a sedimentary rock would not accurately reflect the age of the rock. Why?

Answer: The mineral grains in sedimentary rocks are "recycled" from the weathering of previously existing igneous or metamorphic rocks, forming sediments that are transported, deposited, and cemented into sedimentary rocks. The radiometric ages of the mineral grains reflect the age of the original igneous or metamorphic rock, rather than the time when the grains were deposited and cemented.

One of the most commonly used radioactive isotopes, carbon-14, raises a special challenge because its daughter, nitrogen-14, is the common nitrogen isotope in the Earth's atmosphere. Atmospheric contamination of samples is almost impossible to avoid. Even so, special circumstances let us use carbon-14 as a rock clock. Here is how it is done: Carbon-14, which has a half-life of 5,730 years, is uniformly distributed throughout the atmosphere. While an organism is alive, it will continuously take in carbon from the atmosphere and so will contain the same proportion of carbon-14 as is present in the atmosphere. At the death of the organism this balance ends, because replenishment of carbon-14 by biological processes such as feeding, breathing, and photosynthesis ceases. The amount of carbon-14 in dead tissues continuously decreases by radioactive decay. Radiocarbon dating of a sample requires only a determination of the amount of carbon-14 that remains; measurement of the daughter isotope, nitrogen-14, is not necessary.

Because of its application to organisms (in dating fossil wood, charcoal, peat, bone, and shell material) and its short half-life, radiocarbon dating is enormously valuable in establishing dates for prehistoric human remains and recently extinct animals. In this way, it is of extreme importance in archaeology. It is also of great value in dating the most recent part of geologic history, particularly the latest glacial age. For example, radiocarbon dates have been obtained for samples of wood taken from trees killed by the advance of ice sheets. They show that the ice reached its greatest extent in the Ohio-Indiana-Illinois region about 18,000 to 21,000 years ago.

Carbon-14 and other short half-life radioactive isotopes are not very useful for determining the ages of rocks that are older than about 100 million years. The parent atoms of short half-life isotopes will have decayed during the millions of years since the formation of the rock; if any parent atoms remain, the number will be vanishingly small and extremely difficult to measure analytically. For this reason, radioactive isotopes with much longer half-lives are used to determine the numerical ages of older rocks and fossils, as shown in Table 3.1. Commonly used isotopes include uranium-238 and other isotopes of uranium, potassium-40, and rubidium-87. Similarly, isotopes with long half-lives are not useful for the radiometric dating of very young materials, because so little of the parent isotope will have decayed that the daughter atoms will be exceedingly difficult to detect by analytical methods.

The composition of the material that is to be dated also plays a role in the selection of isotopes. For example, basalt typically does not contain significant amounts of either uranium or potassium, whereas granite commonly contains both of these elements; therefore, uranium and potassium isotopes are generally more useful for dating granitic rocks than for dating basaltic rocks.

Table 3.1 Some Isotopes Used in Radiometric Dating

Parent	Daughter	Half-Life (Years)	Dating Range (Years)
Uranium-238	Lead-206	4.5 billion	10.0 million to 4.6 billion
Uranium-235	Lead-207	710.0 million	10.0 million to 4.6 billion
Thorium-232	Lead-208	14.0 billion	10.0 million to 4.6 billion
Potassium-40	Argon-40	1.3 billion	50,000 to 4.6 billion
Rubidium-87	Strontium-87	4.7 billion	10.0 million to 4.6 billion
Carbon-14	Nitrogen-14	5,730	100 to 70,000

As geologists worked out the geologic column, they discovered that many layers of lava and volcanic ash are interspersed with the sedimentary strata. Using radiometric dating techniques, it is possible to determine the numerical ages of the lavas and volcanic ash layers, and thereby to bracket the ages of the sedimentary strata. Through a combination of geologic correlation and radiometric dating methods, twentieth-century scientists have been able to fit a scale of numerical time to the geologic column worked out on the basis of relative ages in the nineteenth century (see Figure 3.3). The scale is being continuously refined as new radiometric dates become available.

What is the half-life of carbon-14?

Answer: 5,730 years.

8 MAGNETIC POLARITY REVERSAL DATING

Another method that can be used to measure numerical time involves paleomagnetism, the ancient magnetism preserved in rocks. You learned about paleomagnetism in chapter 1 in the context of seafloor spreading.

Recall that some igneous rocks—particularly rocks that contain the mineral magnetite and other iron-bearing minerals—are magnetic. To acquire magnetism, a crystallizing mineral in a lava or magma must cool below a certain temperature, called the **Curie point.** The Curie point for each mineral is different; magnetite, for example, has a Curie temperature of 580°C (1,076°F). At temperatures above the Curie point, the individual magnetic fields of all the iron atoms within the mineral are randomly oriented. As the mineral solidifies and cools through the Curie point, the magnetic fields of the iron atoms all line up in the same direction. They align themselves with the north-south directionality (the polarity) of the Earth's magnetic field at that time, and each of the mineral grains becomes a tiny magnet. Grains of magnetite locked in an igneous rock cannot move and reorient themselves the way a freely swinging bar magnet can. As long as that rock lasts, therefore (until it is destroyed by weathering or metamorphism), it will carry a record of the polarity of the Earth's magnetic field at the moment it cooled below its Curie point.

Figure 3.5

Sedimentary rocks also acquire weak magnetism through the orientation of magnetic grains during sedimentation. As grains settle through ocean or lake water, or even as dust particles settle through the air, any magnetite particles present will act as freely swinging magnets and orient themselves parallel to the magnetic lines of force caused by the Earth's magnetic field (Figure 3.5). Once locked into a sediment, the grains make the rock weakly magnetic.

Sometimes a rock's paleomagnetic polarity is normal, that is, the same as the Earth's present-day polarity. Sometimes it is reversed, that is, opposite to the Earth's present-day polarity. In chapter 1 you learned that every now and then—every half-million years or so—the Earth's magnetic field reverses its polarity: the north magnetic pole becomes the south magnetic pole and the south magnetic pole becomes the north. Scientists don't understand exactly how or why these **magnetic reversals** happen. The important thing for our discussion is that the rock retains the magnetic polarity of the Earth at the time that the rock cooled. From the study of paleomagnetism in lavas, many magnetic reversals have been discovered (Figure 3.6). The ages of lavas, as you know, can be determined using radiometric dating. Through combined radiometric dating and magnetic polarity measurements, scientists have been able to establish a time scale of magnetic polarity reversals dating back to the Jurassic period. Earlier reversals are the topic of ongoing research.

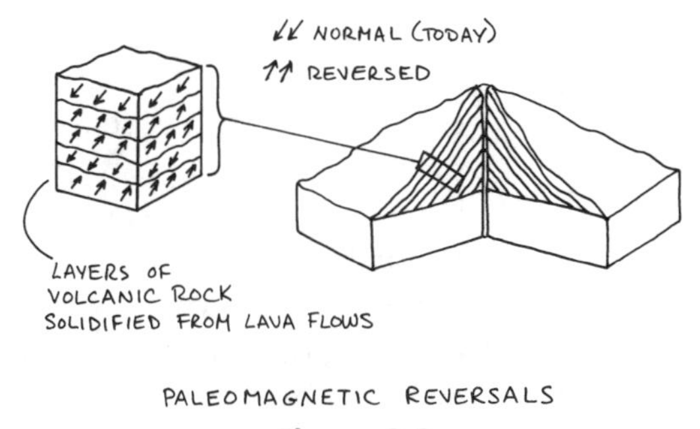

Figure 3.6

The use of magnetic reversals for geologic dating differs from other dating methods, because one magnetic reversal looks just like any other in the rock record. When evidence of a magnetic reversal is found in a sequence of rocks, the challenge is to figure out which one of the many reversals it actually is. When a continuous record of reversals can be found, starting with the present, it is simply a matter of counting backward. Otherwise, it may be necessary to match patterns of reversals, such as two quick reversals followed by a long period of normal polarity. If a specific reversal sequence can be identified, its position in the chronologic sequence of magnetic reversals can be pinpointed.

Paleomagnetism is a very important dating technique for sedimentary rocks.

When fossils are present, an approximate age can be assigned. Knowing the approximate age, geologists can determine the exact age from the magnetic reversals by comparison with the magnetic reversal time scale. Sediment cores recovered from the seafloor can be dated very accurately using a combination of fossils and magnetic reversals. The measurements are so good that magnetic reversals can even provide an accurate way to measure rates of sedimentation in the world's oceans.

Why do reversals of the Earth's magnetic field occur every half-million years or so?

Answer: The phenomenon of magnetic reversals is not fully understood.

9 HOW OLD IS THE EARTH?

The ability to determine the numerical ages of rocks has changed the way we think about the world. But how can we determine the age of the Earth itself? The earliest rocks preserved on the Earth come from the great assemblage of metamorphic and igneous rocks formed during Precambrian time. The oldest radiometric dates, about 4.4 billion years, have been obtained from individual mineral grains in sedimentary rocks from Australia. These grains—of the mineral zircon—show evidence of having experienced wet partial melting during the formation of a magma, which then crystallized into an igneous rock, which in turn was weathered, eroded, and redeposited, eventually to be incorporated into a sedimentary rock. Dates almost as old—4.0 billion years—have been obtained from granitic igneous rocks from Canada. The existence of such ancient rocks proves that continental crust was present 4.0 billion years ago, while the 4.4-billion-year-old mineral grains prove that the cycle of weathering, erosion, deposition, and cementation was operating then. Because we see wet melting and ancient sediment that was transported by water, we know that there must have been water on the surface of the Earth at the time the sediments were deposited.

These ancient rocks are all from the Archean eon. Recall that the Hadean eon predates the Archean. No rocks that might provide radiometric dates are preserved from early Hadean time—none that we have yet found, at any rate. How long did the Hadean eon last, and, therefore, how much older might our planet be? Strong evidence from astronomy suggests that Earth formed at the same time as the Moon, the other planets, and meteorites. Through radiometric dating, it has been possible to determine the ages of meteorites and of Moon rocks brought back by astronauts. The ages of the most primitive of these objects cluster closely around 4.56 billion years. By inference, the time of formation of Earth, and indeed of all the other planets and meteorites in the solar system, is believed to be 4.56 billion years ago.

What is the age of the oldest rock found on the Earth so far, and where is it from?

____

Answer: 4.0 billion years; Canada (older ages have been obtained from individual mineral grains).

Now take some time to review the material in this chapter, and test your understanding of the material by trying out the Self-Test.

SELF-TEST

These questions are designed to help you assess how well you have learned the concepts presented in chapter 3. The answers are given at the end.

1. The principle of crosscutting relationships says that ___________.
   a. a rock unit is older than a feature that disrupts it, such as a fracture
   b. waterborne sediments are deposited in horizontal layers
   c. disrupted sediments are older than sediments that occur in horizontal layers
   d. a sedimentary rock unit is younger than the layer below it, and older than the layers that overlie it
2. Carbon-14 dating is most useful for establishing the age of relatively young materials because it ___________.
   a. was not abundant on the Earth prior to 100 million years ago
   b. has a short half-life
   c. decays very slowly
   d. is found only in recent biologic organisms
3. The three eras that make up the Phanerozoic eon are the ___________.
   a. Hadean, Archean, and Proterozoic
   b. Paleozoic, Proterozoic, and Pleistocene
   c. Triassic, Jurassic, and Cambrian
   d. Paleozoic, Mesozoic, and Cenozoic
4. The oldest known rock on the Earth is 4.0 billion years old, but there are individual mineral grains that are even older. (T or F)
5. The geologic column has four major divisions, which are called eras. (T or F)
6. An isotope that is undergoing radioactive decay is called a(n) ___________, and an isotope that forms as a result of radioactive decay is called a(n) ___________.
7. The periods of the geologic time scale are subdivided into smaller units of time called ___________.

8. The dinosaurs were dominant during the __________ era.

9. How did William Smith's work lead to the use of stratigraphic correlation and eventually the development of the geologic column?

____

____

____

10. What is an isotope?

____

____

____

11. Why is the half-life of a radioactive isotope important in making a choice of which isotope to use in determining the radiometric age of a mineral or rock?

____

____

____

12. What is the Curie point, and why is it important for magnetic reversal dating?

____

____

____

13. If an atom loses an electron, what changes: the atomic number, the mass number, both, or neither one? If an atom loses a neutron, what changes: the atomic number, the mass number, both, or neither one?

____

ANSWERS

1. a
2. b
3. d
4. T
5. F
6. parent; daughter

7. epochs
8. Mesozoic
9. Based on Smith's discovery that each stratum contained a specific group of fossils, geologists were able to show that stratigraphic successions can be correlated from one locality to another. Soon, worldwide correlations were made. This provided a means of equating the ages of rocks from different localities, which eventually led to the establishment of the geologic column.
10. Different forms of the same element. Isotopes of the same element have the same atomic number (same number of protons in the nucleus) but different mass numbers (different numbers of protons + neutrons in the nucleus).
11. If the half-life of the radioactive isotope is very short and you are trying to measure the age of a mineral or a rock that is very old, then the parent isotope will be completely or nearly gone from the rock, having decayed away to vanishingly small amounts long ago. If the half-life of the radioactive isotope is very long and you are trying to measure the age of a mineral or a rock that is very young, so little of the radioactive material will have decayed that the daughter atoms will be difficult to detect and measure, even with sensitive analytical techniques.
12. The Curie point is the temperature below which iron-bearing mineral grains in a cooling lava or magma acquire permanent magnetism. It is important for magnetic dating because the mineral grains retain the same polarity as the Earth's magnetic field at the time when the grains cooled through the Curie point. Thus, a sequence of lavas that erupted over a period of time will record successive reversals in the Earth's magnetic field. These reversals have been correlated worldwide and specific dates have been assigned to them through radiometric dating.
13. When losing an electron, neither one changes. When losing a neutron, the mass number changes.

KEY WORDS

Archean eon
Cenozoic era
Curie point
eon
epoch
era
fossil
geochronology
geologic column
Hadean eon
half-life
isotope
magnetic reversal
Mesozoic era
numerical (absolute) age
paleontology
Paleozoic era
period
Phanerozoic eon
Precambrian
principle of crosscutting relationships
principle of original horizontality

- principle of stratigraphic superposition
- Proterozoic eon
- radioactive decay
- radioactivity
- radiometric dating
- relative age
- strata (singular: stratum)
- stratigraphic correlation
- stratigraphy
- unconformity
- uniformitarianism

4 Earthquakes and the Inside of the Earth

We rode on a sea of mountains and jungles, sinking in rubble and drowning in the foam of wood and rock. The Earth was boiling under our feet.

—Anonymous survivor of a 1773 earthquake in Guatemala

OBJECTIVES

In this chapter you will learn

- how scientists study the inside of the Earth;
- what causes earthquakes;
- why earthquakes are valuable tools for learning about the Earth's interior;
- how the different layers and boundaries inside the Earth have been recognized.

1 INSIDE THE EARTH

We know a lot about the Earth's surface because we can move around on it and collect samples for study. However, we know less about the Earth's interior because it is mostly inaccessible. We *do* know that the Earth—like the other terrestrial planets—is a differentiated body, which means that early in its history it separated into a core, a mantle, and a crust, each with different chemical compositions and physical characteristics. But what exactly are the characteristics of the deeper layers, and how were they discovered? How did scientists realize that what happens deep inside the Earth influences what happens on the surface? This chapter addresses these questions.

There are two ways of studying something scientifically. You can study it by direct sampling; that is, collect samples, examine them, and analyze them in a laboratory. Or you can study it "remotely," using techniques that enable you to collect information without actually coming into contact with the object. By comparison, medical techniques such as X rays allow doctors to study the inside of the body without having direct access to it. Geologists use both direct and remote techniques to study the inside of the Earth.

What is the difference between "direct" and "remote" methods of studying something?

____

____

Answer: In direct studies, samples are collected and examined or analyzed in a laboratory. In remote studies, scientists have no direct access, but must analyze the object from a distance.

2 SAMPLING THE EARTH'S INTERIOR

If you were given the task of trying to collect samples from inside the Earth, you might try to drill a very deep hole and collect samples from the bottom of the hole. Indeed, scientists have tried this approach. The deepest hole ever drilled reached a depth of almost 12 km (7.5 mi) in the Kola Peninsula, Russia. Recall from chapter 1 that the average thickness of oceanic crust is about 8 km (5 mi). A 12-km-deep hole sounds just about right for sampling the top of the mantle—or does it?

If you tried to drill a hole through thin oceanic crust to the mantle, you would quickly encounter high temperatures that would destroy your equipment. This is because areas where the crust is thin tend to have high heat flow. Another problem is that oceanic crust is deep underwater, which makes drilling difficult. The only place where the rocks are both accessible and cool enough for drilling to great depths is where the crust is very thick—that is, the continents (average thickness 45 km, or about 30 mi). The hole in the Kola Peninsula went through 12 km of continental crust and never even came *close* to reaching the mantle. Drilling has yielded much interesting information about the composition of the crust, the distribution of heat, the flow of fluids through the crust, and how materials change with depth. However, it isn't useful for sampling the mantle—at least not with today's drilling technologies.

Another way to get samples from deep within the Earth is to wait for them to be delivered to the surface by natural geologic processes. Molten rock (magma) forms deep in the Earth. When the magma rises to the surface, fragments of the surrounding rock may be carried along. We call these fragments **xenoliths,** from the Greek words *xenos* ("foreigner") and *lithos* ("stone"). Of particular interest is a rock type called kimberlite, which forms from magma that originates at depths of

150 to 300 km (about 100 to 200 mi)—deeper than even the thickest portions of the crust. Kimberlite magmas rip fragments of mantle rock from their source region and carry them to the surface (Figure 4.1). These xenoliths contain minerals, such as diamonds, that can only form under conditions of very high pressure and temperature, correlating to depth ranges of 150 to 300 km.

What is a xenolith?

Answer: A sample of rock carried to the surface by magma from deep within the Earth.

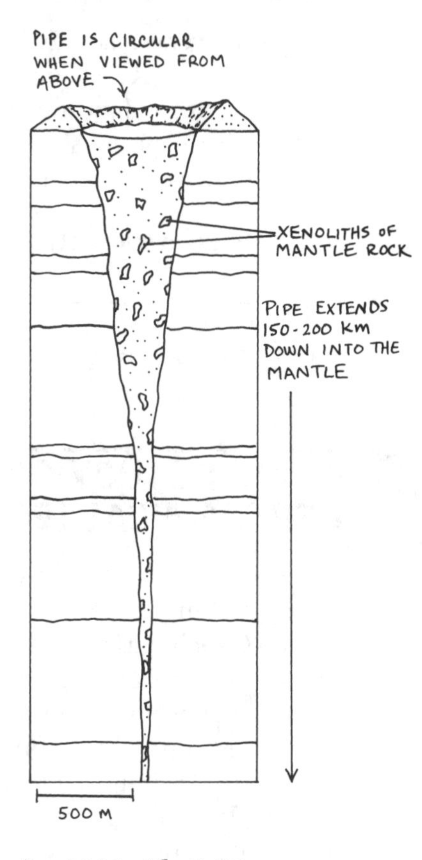

Figure 4.1

3 REMOTE STUDY OF THE EARTH'S INTERIOR: ASTRONOMICAL TECHNIQUES

Scientists use the basic tools of astronomy, chemistry, and physics to learn about the Earth's interior without actually sampling it. We begin by using the same techniques that are used to study the interiors of other planets. First we determine the planet's mass by observing its gravitational influence on other planets and satellites, then applying the physical laws that govern planetary motion. We also need to know the diameter of the planet. This was challenging for early geoscientists, who painstakingly calculated the Earth's diameter from thousands of careful surveying measurements. Today it is measured quickly and accurately by satellites and global positioning systems. Once we know the dimensions of the planet, it is relatively simple to figure out the volume as long as we know its exact shape—another problem that seems trivial now but was very challenging for early geoscientists. (In fact, we now know that the Earth is not exactly spherical; it bulges out slightly at the equator.) Once we know the mass and volume of the planet, we can figure out its overall density (mass ÷ volume).

These measurements and calculations tell us the size, mass, and density of the planet, but what do they reveal about its interior? For one thing, they can tell us whether material is distributed evenly throughout the planet. Rocks at the surface are very light (low-density) compared to the planet as a whole. Surface rocks have an average density of about 2.8 g/cm^3, whereas the Earth's overall density is 5.5 g/cm^3. (Recall from chapter 2, for comparison, that water has a density of 1.0 g/cm^3, and most common minerals have densities in the range of 2.5 to 3.0 g/cm^3.) For the planet as a whole to have such a high density, there must be a concentration of denser material inside the planet.

How do scientists measure the Earth's diameter?

Answer: Using satellites and global positioning systems.

4 REMOTE STUDY OF THE EARTH'S INTERIOR: GEOCHEMICAL TECHNIQUES

Scientists also use chemistry to study the Earth's interior. The use of chemical techniques to study the Earth's materials and processes is called **geochemistry.** Geochemists conduct experiments in which rocks and minerals are subjected to extreme pressures and temperatures. In this way, they can learn what types of chemical compounds are formed and what kinds of processes may occur under the conditions found at great depths in the Earth.

An important geochemical technique involves the study of meteorites, samples of extraterrestrial materials that have fallen to Earth. Most meteorites formed at about the same time and, in some cases, the same part of the solar system as the Earth. Some meteorites are chemically "primitive"; that is, they have never been affected by melting, metamorphism, or differentiation since the time of their formation. The Earth, in contrast, has been greatly changed by various geologic processes, and its chemical elements have been redistributed. If we examine primitive meteorites we can get an idea of what the Earth as a whole must have been like prior to the redistribution of its elements. Other meteorites are fragments of planetary objects that differentiated into core, mantle, and crust, just as the Earth did. Some of these meteorites—notably the "irons," which are composed of iron and nickel metal—are thought to be similar to the Earth's core. A core of the same composition as an iron meteorite would provide just the right concentration of dense material at the center of the Earth to yield an overall density of 5.5 g/cm^3.

What is a "primitive" meteorite?

Answer: A meteorite that has not been altered by geologic processes such as melting, metamorphism, or differentiation since its formation.

5 REMOTE STUDY OF THE EARTH'S INTERIOR: GEOPHYSICAL TECHNIQUES

The application of physics to the study of the Earth is called **geophysics.** One geophysical technique involves measuring variations in gravity. Gravity is the attractional force that causes a downward pull on any object at the Earth's surface. It is measured with a gravimeter, which is basically a weight suspended from a spring attached to sensitive measuring devices. The stronger the pull of gravity, the more the weight is pulled down and the more it stretches the spring. The stretching of the spring provides a measure of the local pull of gravity.

How can gravity measurements tell us about the material inside the Earth? If the rocks between the surface and the center of the Earth were the same everywhere, the force of gravity should be the same for every point on the surface. However, gravity measurements reveal variations, called **gravity anomalies,** caused by underlying bodies of rock with differing densities (Figure 4.2). (Gravity *feels* the

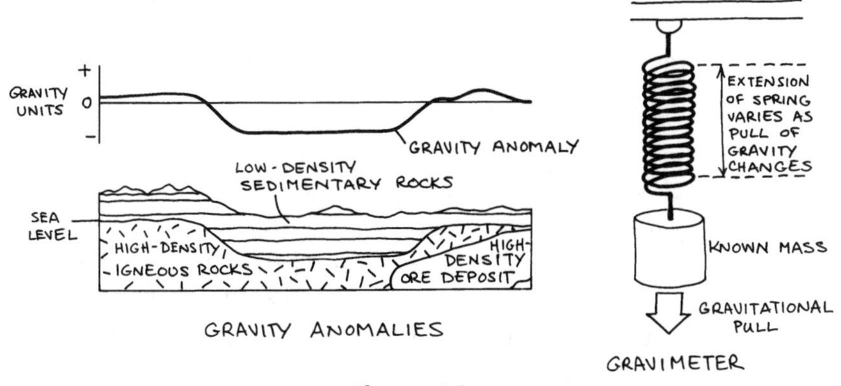

Figure 4.2

same to us everywhere on the Earth because the variations are so small that we cannot detect them without sensitive instruments.) Because the pull of gravity depends on mass, a heavy (dense) mass will cause a greater pull than a light (less dense) mass. A greater-than-average pull is called a "positive" anomaly; a less-than-average pull is called a "negative" anomaly. By making gravity measurements over the Earth's entire surface, we can gain information about the distribution of dense and less dense materials underground. One of the most important things this type of measurement has revealed is that mountains have roots that extend to great depths.

Magnetism is another physical force that can provide information about the inside of the Earth. The Earth has a magnetic field (refer to Figure 3.5, page 61), so there must be something inside that is generating the magnetism. We could think of the Earth as having a huge dipole magnet (a bar magnet with north and south poles) at its center. The problem is that bar magnets lose their magnetism at temperatures above about 500°C (900°F, their Curie point), and we know that the temperature deep inside the Earth is *much* higher than this. So, how does the Earth generate its magnetic field?

The most widely accepted explanation is the **dynamo hypothesis.** A conducting material moving through the magnetic field of a bar magnet generates electricity. If the bar magnet is replaced by a coil of wire, the electric current will continue to sustain the magnetic field. Through a mechanism that is similar but more complex, the movement of an electrically conducting liquid inside a planet can generate and sustain a magnetic field. Planets with strong magnetic fields are thought to have cores that are liquid (freely moving) and metallic or otherwise conductive. The Earth's core is mainly metallic iron; movement within the liquid part of the iron core generates the magnetic field.

How do we know the Earth's magnetic field is not generated by a simple bar magnet?

Answer: A bar magnet would lose its magnetism at the very high temperatures that characterize the Earth's interior.

The single most important geophysical technique scientists use to learn about the Earth's interior is the study of vibrational waves generated by earthquakes. Earthquake vibrations are sort of like the X rays a doctor uses to study the inside of a human body; by examining how earthquake waves travel through the deep, inaccessible layers of the Earth, we can learn about the physical characteristics of those layers. Let's begin by examining earthquakes and what causes them. Then we can evaluate the information earthquake waves provide about the interior of the Earth.

6 WHAT CAUSES EARTHQUAKES?

Earthquakes occur in specific tectonic settings, primarily along plate boundaries. This close connection suggests that the mechanisms that cause earthquakes must be related to those that drive plate motion. But what, exactly, causes earthquakes?

Earthquakes are caused by the sudden movement of strained blocks of the Earth's crust. Tectonic forces produce stress, or directional pressure, which causes large blocks of rock to break along a large fracture, or **fault.** When the blocks grind past one another along the fault, an earthquake results. If the movement occurs near the Earth's surface, it may disrupt and displace surface features. Movement along a fault and the resulting displacement of surface features can be horizontal or vertical (sometimes both). The largest abrupt vertical displacement ever observed occurred in 1899 at Yakutat Bay, Alaska, when a stretch of the Alaskan shore was suddenly lifted 15 m (50 ft) above sea level during a major earthquake. In other cases, fault blocks creep past each other slowly but continuously. (You will learn more about the various types of faults in chapter 8.)

The most widely accepted explanation for the origin of earthquakes is the **elastic rebound theory.** It is based on the mechanics of elastic deformation of rocks, that is, reversible changes in the volume or shape of a rock that is subjected to stress (more on this in chapter 8, as well). When the stress is removed, the elastically deformed material snaps back to its original size and shape, causing an earthquake.

You can demonstrate the storage of energy in an elastically deformed material with a steel spring, a long metal ruler, or a heavy rubber band. When you compress the spring, bend the ruler, or stretch the rubber band, the material undergoes strain in the form of elastic deformation. If you suddenly release the material, it bounces back to its original shape, releasing the built-up energy with a "twang."

Elastic rebound theory states that energy can be stored in bodies of rock when they are subjected to stress along a fault plate. Eventually, the increasing stress along the fault is sufficient to overcome the friction between the blocks. The blocks slip, the stored energy is suddenly released in the form of an earthquake, and the rocks rebound to assume their original shapes (Figure 4.3).

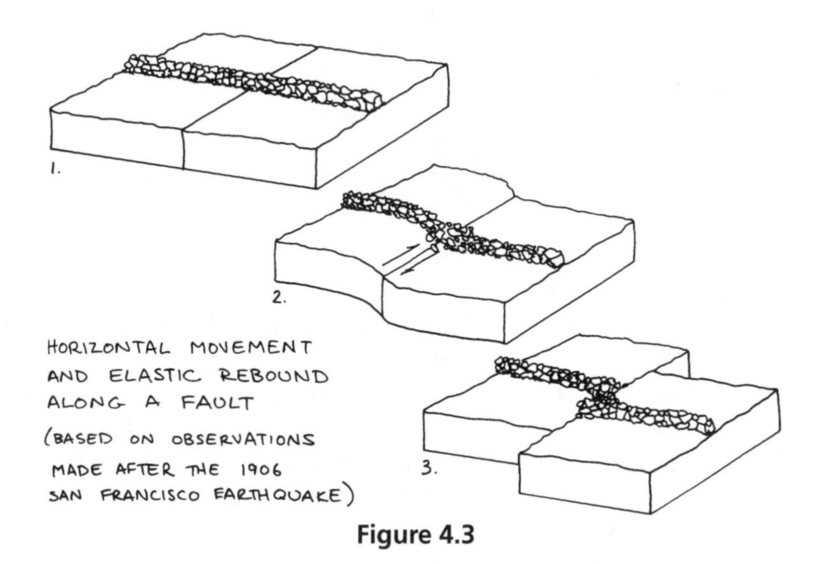

Figure 4.3

What is elastic deformation?

Answer: A reversible change in the volume or shape of a rock that is subjected to stress.

7 HOW EARTHQUAKES ARE STUDIED

The study of earthquakes is called **seismology,** from the Greek word *seismos,* "earthquake." Scientists who study earthquakes are **seismologists,** and the devices used to record earthquake vibrations are **seismographs** (Figure 4.4). A seismo-

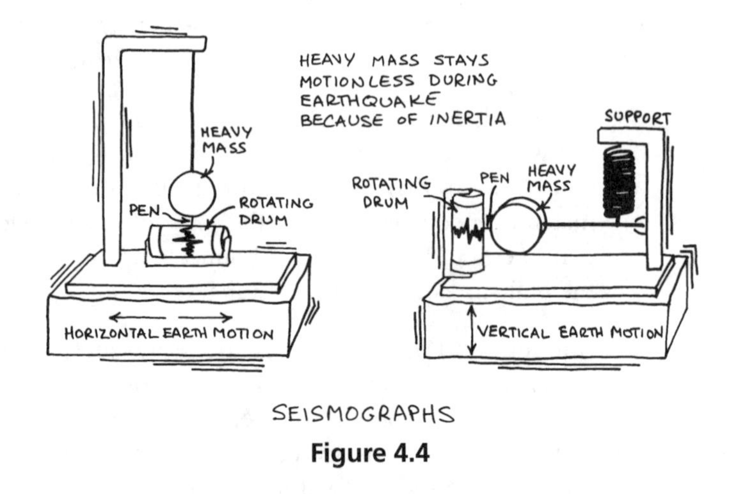

Figure 4.4

graph stands on the Earth's vibrating surface, so it will vibrate along with that surface. This makes measurement difficult because there is no fixed frame of reference against which to make measurements. The problem is the same one that a sailor in a small boat would face when attempting to measure waves at sea. To overcome this frame-of-reference problem, most seismographs utilize the property of **inertia,** which is the resistance of a large stationary mass to sudden movement.

If you suspend a heavy mass, such as a block of iron, from a spring and suddenly lift the upper end of the spring, you will notice that the block remains almost stationary because of inertia, while the spring stretches upward.

To measure vertical ground movement, a heavy mass is suspended from a spring and the spring is connected to a support, which in turn is connected to the ground. When the ground vibrates, the support moves and the spring expands and contracts, but the mass remains almost stationary. The distance between the ground and the mass can be used to sense movement of the ground surface. Horizontal movement can be measured by suspending a heavy mass from a string to make a pendulum. Because of inertia, the mass does not keep up with the motion of the ground. The difference between the movement of the pendulum and that of the ground serves as a measure of ground motion. Modern seismographs are incredibly sensitive because the movements they detect are measured optically and amplified electronically. Vibrational movements as tiny as 100 millionths of a centimeter can be detected.

What is a seismologist?

Answer: A scientist who studies earthquakes.

8 SEISMIC WAVES

Most of the energy released by an earthquake is transmitted to other parts of the Earth. The released energy travels outward in the form of vibrational waves from the earthquake's source, or **focus** (plural: **foci**) (Figure 4.5). These waves, called **seismic waves,** are elastic disturbances. This means that the rocks through which they pass return to their original shapes after the waves have passed; the waves therefore must be measured while the rock is still vibrating. For this reason, a network of continuously recording seismograph stations has been installed around the world. When an earthquake occurs, seismic waves are recorded by many seismographs and instantly evaluated by computers. Records obtained in this way are called **seismograms.**

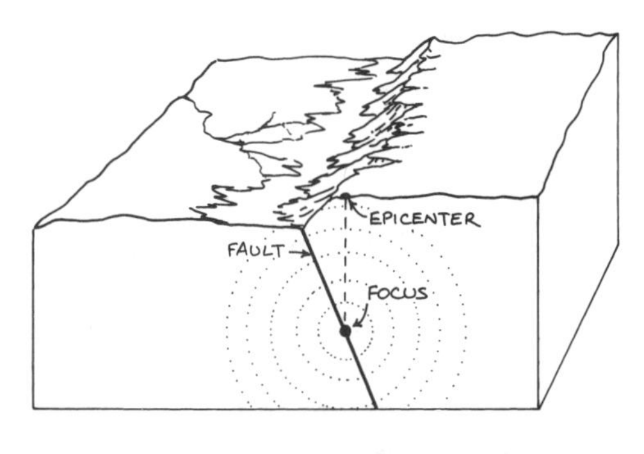

Figure 4.5

There are two types of seismic waves. **Body waves** travel outward from the focus through the interior of the Earth. **Surface waves** are restricted to the Earth's surface.

Compressional waves, one of two types of seismic body waves, consist of alternating pulses of compression and expansion acting in the direction in which the wave is traveling (Figure 4.6). Sound waves are also compressional waves. Compressional waves can pass through solids, liquids, and gases. They have the greatest velocity of all seismic waves—6 kilometers/second (km/s) (almost 4 miles/second [mi/s]) is a typical velocity for compressional seismic waves through the upper crust. They are the first waves to be recorded after an earthquake, so they are called **P** (for "primary") **waves.**

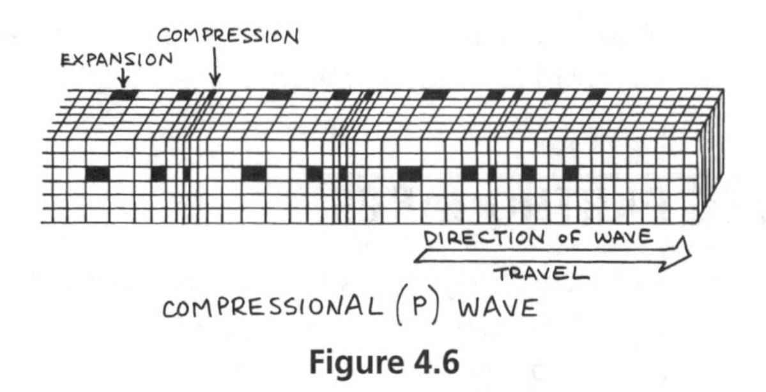

Figure 4.6

Shear waves, the other type of body wave, travel through materials in an alternating series of sidewise movements (Figure 4.7). Shearing involves changing

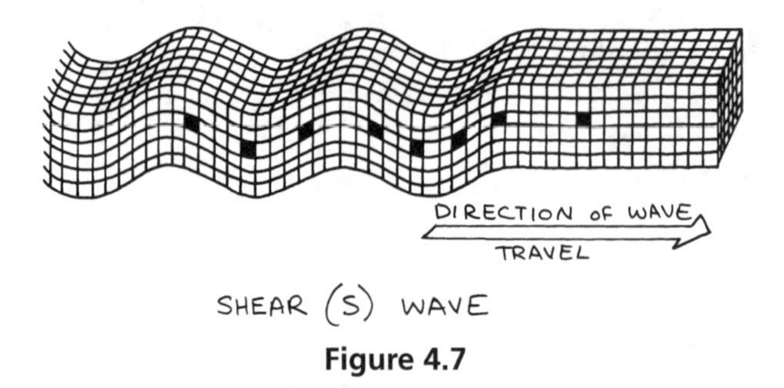

Figure 4.7

the shape of an object. Solids have elastic characteristics that provide a restoring force for recovery from shearing, but liquids and gases lack these elastic characteristics. Therefore, shear waves cannot be transmitted through liquids or gases. This is very important, as you will soon see. A typical velocity for shear waves in the upper crust is 3.5 km/s (more than 2 mi/s). Shear waves are slower than P waves, so they reach a seismograph some time after the arrival of P waves from the same earthquake. For this reason, they are called **S** (for "secondary") **waves.**

Surface waves travel along or near the surface of the Earth. They travel more slowly than P and S waves, and they pass around the Earth rather than through it. Thus, surface waves are the last to be detected by a seismograph. It is important for planners and builders to understand surface waves, because they cause much of the ground shaking that damages buildings and infrastructure (roads, pipes, sewers, etc.) during large earthquakes. Underground bomb blasts, such as those sometimes used for nuclear bomb testing, generate more surface waves than earthquakes do; this difference helps scientists detect bomb blasts, an important part of nuclear test ban treaty verifications.

What is the difference between a seismograph and a seismogram?

____

____

Answer: A seismograph is the instrument used to measure earthquake vibrations. A seismogram is a record (on paper) of earthquake vibrations.

9 LOCATING EARTHQUAKES

If an earthquake's body waves have been recorded by three or more seismographs, the location of its **epicenter**—the point on the Earth's surface directly above the focus (Figure 4.5)—can be determined through simple graphical triangulation.

The first step is to find out how far each seismograph is from the epicenter. This is done by comparing the arrival times of P and S waves (Figure 4.8). The greater the distance traveled by the waves, the greater the difference between their arrival times at the seismic station. In other words, the farther the waves travel, the more the S waves lag behind the P waves.

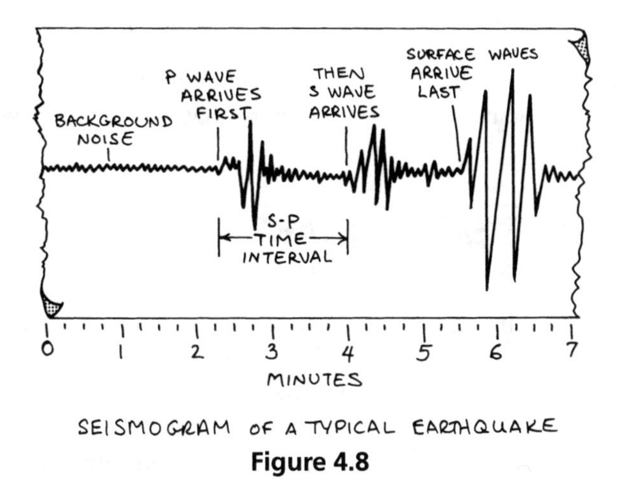

SEISMOGRAM OF A TYPICAL EARTHQUAKE

Figure 4.8

After determining the distance from the seismograph to the epicenter, the seismologist draws a circle on a map, with the seismic station at the center of the circle (Figure 4.9). The distance from the seismograph to the epicenter is the radius of the circle. It is a circle because the seismologist knows only the *distance* to the epicenter—usually there is no easy way to know the *direction* from which

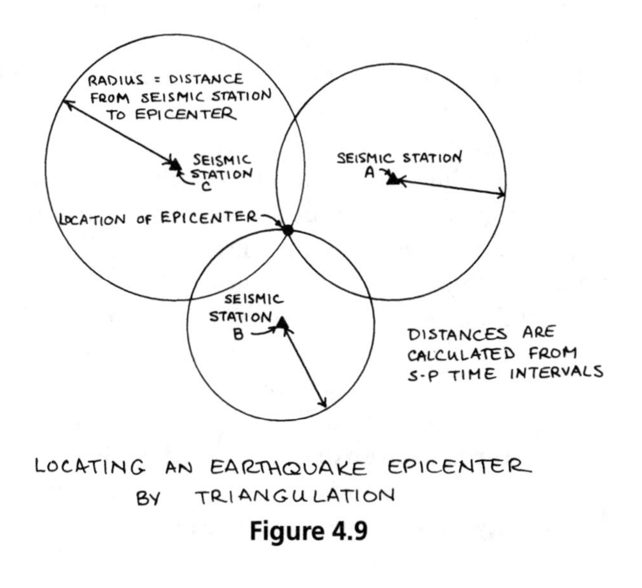

LOCATING AN EARTHQUAKE EPICENTER BY TRIANGULATION

Figure 4.9

the waves have come. Similar information is plotted for three or more seismographs. The intersection of the circles is the location of the epicenter.

What is the difference between the epicenter and the focus of an earthquake?

Answer: The focus is the point of origin of the earthquake's release of energy. The epicenter is the point on the Earth's surface that is directly above the focus.

10 MEASURING EARTHQUAKES

Scientists are also interested in measuring and quantifying the strength of earthquakes. The most widely used measures of earthquake intensity are the Richter magnitude, the modified Mercalli intensity, and the moment magnitude.

The **Richter magnitude,** named after seismologist Charles Richter, who developed it in 1935, is calculated from the amplitudes (wave heights) of P and S waves recorded on a seismogram. The Richter magnitude scale is logarithmic; this means that every increase of one on the scale corresponds to a tenfold increase in the amplitude of the wave signal. Thus, a magnitude 6 earthquake has a seismic wave amplitude 10 times larger than that of a magnitude 5 quake. A magnitude 7 earthquake has an amplitude 100 times larger than that of a magnitude 5 quake (a tenfold increase for each step, 10×10). Richter magnitudes are corrected for distance from the epicenter. This means that the Richter magnitude for a given earthquake is the same whether you are standing at the epicenter, where the effects of the earthquake would feel strongest, or thousands of kilometers away, where the effects would feel less intense.

Each step in the Richter scale corresponds to a tenfold increase in the amplitude of the seismic wave signal. However, the amount of energy actually released during the course of an earthquake is closer to a thirtyfold increase for each step in the scale. Thus, a magnitude 6 earthquake releases approximately 900 times as much energy as a magnitude 4 quake (a thirtyfold increase for each step in the scale, $30 \times 30 = 900$). A magnitude 8 earthquake may release almost a million times as much energy as a magnitude 4 quake ($30 \times 30 \times 30 \times 30 = 810{,}000$). So, even though large earthquakes happen infrequently, a single, very large quake can release as much stored energy as many thousands of smaller ones. The largest earthquakes on record have had Richter magnitudes of about 8.6; an example is the Alaskan "Good Friday" earthquake of 1964.

The **Mercalli intensity,** developed in the late 1800s and later modified, is based on eyewitness reports and on the extent of damage to buildings. The modified Mercalli intensity scale ranges from I (barely felt) to XII (practically all works of con-

ALASKAN EARTHQUAKE DAMAGE. One of the largest earthquakes ever recorded, the "Good Friday" earthquake of March 27, 1964, caused extensive damage in Anchorage, Alaska. The collapse of Fourth Avenue, shown here, was caused partly by the loss of strength of the soil and sediments as a result of shaking by the earthquake waves. (Courtesy U.S. Army)

struction destroyed or severely damaged), as shown in Table 4.1. The Mercalli intensity of an earthquake varies with distance from the epicenter. A single earthquake could have a Mercalli intensity of IX or X near the epicenter, where the intensity is greatest, whereas a few hundred kilometers away its intensity would be only I.

The Mercalli scale is particularly useful in the study of earthquakes that occurred before the development of modern seismologic equipment. For example, the exact magnitudes of a series of devastating earthquakes that struck New Madrid, Missouri, in 1811–1812 are unknown. However, historical eyewitness accounts suggest that all three earthquakes had Mercalli intensities of XI near the epicenter. Combining this with information about the nature of the bedrock and how far away the effects were felt, researchers estimate that the quakes had Richter magnitudes of approximately 7.8.

The **seismic moment magnitude** is an expression of the strength of an earthquake based on observations of the area over which the fault ruptured and the average ground displacement. Richter magnitude calculations are based on the concept that earthquake foci are points; therefore, the Richter scale is best suited as a measure of the strength of an earthquake in which energy is released from a relatively small volume of rock. In contrast, the calculation of moment magnitude takes account of the fact that energy may be released from a large area along a

Table 4.1 Earthquake Magnitudes* and Characteristic Damage

Richter Magnitude	Mercalli Magnitude	Number per Year	Characteristic Effects
<3.4	I	800,000	Recorded only by seismographs
3.5–4.2	II and III	30,000	Felt by some people
4.3–4.8	IV	4,800	Felt by many people; windows rattle
4.9–5.4	V	1,400	Felt by everyone; dishes break, doors swing
5.5–6.1	VI and VII	500	Slight building damage; plaster cracks, bricks fall
6.2–6.9	VIII and IX	100	Much building damage; chimneys fall; houses move on foundations
7.0–7.3	X	15	Serious damage; bridges twisted, walls fractured; many masonry buildings collapse
7.4–7.9	XI	4	Great damage; most buildings collapse
>8.0	XII	1 every 5–10 years	Total destruction; ground surface waves are visible

*Note that it is not possible to convert Richter magnitudes into exact equivalents on the Mercalli intensity scale; like apples and oranges, they are not directly comparable.

fault. Seismic moment also accounts for variations in the physical characteristics of Earth materials, which can affect the efficiency with which seismic waves are transmitted. Media reports generally give Richter magnitudes, even though the moment magnitude may be a better indication of the true strength of an earthquake in some cases. For comparison, the moment magnitude of the Alaskan "Good Friday" earthquake of 1964 was 9.2.

An earthquake always has just one Richter magnitude, instead of having a high magnitude near the epicenter and a lower magnitude away from the epicenter. Why?

Answer: The Richter magnitude calculation includes a correction for distance from the epicenter.

11 EARTHQUAKE HAZARDS

Each year hundreds of thousands of earthquakes occur. Fortunately, only a few are large enough, or close enough to major population centers, to cause loss of life. The most disastrous quake in history occurred in Shaanxi, China, in 1556; an estimated 830,000 people died. The worst earthquake disaster of the twentieth century also occurred in China. On July 28, 1976, a 7.8 magnitude quake leveled the city of Tangshan. Hardly a building was left standing, and the few that withstood the first earthquake were destroyed by a second one that struck later the same day. At least 240,000 people were killed. In all, seventeen earthquakes are known to have caused 50,000 or more deaths apiece.

Ground motion is the primary cause of damage during an earthquake. Proper design of buildings and other structures can do much to prevent damage, but in a very strong earthquake even the best-designed buildings may collapse. Where a fault breaks the ground surface, buildings can be split, roads disrupted, and anything that lies on or across the fault broken apart. Major earthquakes are often followed by more (usually smaller) earthquakes called **aftershocks.** The Landers (Los Angeles) earthquake of 1992 triggered major aftershocks at 14 locations, some nearby, others hundreds of kilometers from the original epicenter.

Earthquakes can also cause landslides. In some circumstances, sudden shaking can turn seemingly solid ground into a quicksandlike material. This process, called **liquefaction,** was a major cause of damage during the earthquake that destroyed much of Anchorage, Alaska, in 1964. Fire is another common secondary effect of earthquakes. Ground motion breaks gas lines and loosens electrical wires, often causing fires. If water mains are broken, there may be no water available to put out the fires. In the Great Earthquake that struck San Francisco in 1906, most of the damage was caused by fires that destroyed 521 city blocks in three days. For many years the quake was referred to as "the Great Fire." In 1989, fires caused by the Loma Prieta "World Series" earthquake again ravaged the Marina district of downtown San Francisco.

THE GREAT SAN FRANCISCO EARTHQUAKE OF 1906. This scene of destruction on Nob Hill, in San Francisco, was caused by the Great Earthquake of 1906. Over 400 km (250 mi) of the San Andreas Fault ruptured and moved during the earthquake. Much of the damage was caused by fires in the aftermath of the earthquake; for years, many people referred to the event as "the Great Fire." (Courtesy Library of Congress)

Another secondary effect of earthquakes is seismic sea waves, or **tsunamis** (sometimes mistakenly called "tidal waves," although they have nothing to do with tides). Sub-marine earthquakes are the main cause of tsunamis, which are especially destructive around the Pacific Rim. A well-known example is the tsunami generated by a sub-marine earthquake near Unimak Island, Alaska, in 1946. The wave traveled across the Pacific Ocean at a velocity of 800 kilometers per hour (km/h) (almost 500 miles per hour [mi/h]), striking Hilo, Hawaii, about 4.5 hours later. When it hit, the wave had a crest 18 m (60 ft) higher than the normal high tide. It demolished nearly 500 houses, damaged 1,000 more, and killed 159 people. More recently, four destructive Pacific Rim tsunamis (Nicaragua, 1992; Flores Island, Indonesia, 1992; Hokkaido, Japan, 1993; and southeast Java and Bali, Indonesia, 1994) killed a total of 1,700 people.

What is liquefaction?

____

____

Answer: The process whereby seemingly solid ground, when subjected to sudden shaking, can turn into quicksand.

12 EARTHQUAKE PREDICTION

Charles Richter once said, "Only fools, charlatans, and liars predict earthquakes." Today, seismologists attempt to predict earthquakes using sensitive instruments to monitor seismically active zones. It still is not possible to predict the exact magnitude and time of occurrence of an earthquake; however, scientists' understanding about seismic mechanisms and the tectonic settings in which earthquakes occur has improved greatly since Richter's time. Advances in modern seismology may yet prove him wrong.

Long-term earthquake forecasting—the prediction of a large earthquake years or even decades in advance—is based mainly on understanding the tectonic cycle and the geologic settings in which earthquakes occur. Where earthquakes occur repeatedly, such as along plate boundaries, it is sometimes possible to detect a regular pattern in the recurrence intervals of large quakes. This requires information about seismic activity going back farther than historical records, which can be provided by **paleoseismology,** the study of prehistoric earthquakes.

The primary goal of paleoseismology is to search the stratigraphic record for evidence of major earthquakes and discern the time intervals between them. If the pattern of earthquake recurrence suggests intervals of, say, a century between major quakes, it may be possible to predict when a large quake is due to happen next in that location. Such studies have identified a number of **seismic gaps** around the Pacific Rim. These are places along a fault where a large earthquake has not occurred for a long time, even though tectonic movement is still active. Seismic gaps are considered by seismologists to be the most likely locations for large earthquakes to occur.

Long-term earthquake forecasting has met with reasonable success. Seismologists know where the most hazardous areas are. They can calculate the probability that a large earthquake will occur in a particular area within a given period. And they have a theory of earthquake generation that successfully unites their predictions and observations in the context of plate tectonic theory.

Short-term prediction of earthquakes has been less successful than long-term forecasting, partly because seismic processes are hidden from view, under the ground. Earthquakes are also highly inconsistent; sometimes they give no discernible warning signs at all. Therefore, short-term prediction remains an elusive goal for seismologists. Precise short-term prediction of the time, location, and magnitude of an earthquake could allow authorities to issue an early warning.

Efforts at short-term prediction are based on observations of precursory phenomena—anomalous occurrences that may serve as signals of impending earthquake activity. Most research involves monitoring changes in the physical properties of rocks, such as magnetism and electrical conductivity. Even changes in the level of water in wells or the amount of radon gas in well water might indicate changes in the properties of the underlying rocks. Strange animal behavior, glowing auras, and unusual radio waves have all been reported near the epicenters of large earthquakes.

Tilting or bulging of the ground and changes in elevation are among the most reliable indications that strain energy is building up. Small cracks and fractures that develop in severely strained rock can cause swarms of tiny earthquakes—**foreshocks**—that may be a clue that a big quake is coming. The most famous successful earthquake prediction, in 1975, was based on slow tilting of the land surface, fluctuations in the magnetic field, and numerous foreshocks that preceded a magnitude 7.3 quake in Haicheng, China. Half the city was destroyed, but because authorities had evacuated more than a million people before the quake, only a few hundred were killed.

What is the role of foreshocks in earthquake prediction?

____

____

Answer: Swarms of foreshocks may indicate that rocks at depth are severely strained and beginning to crack and that an earthquake is imminent.

13 HOW EARTHQUAKES REVEAL THE EARTH'S INTERIOR

Seismograms from seismic stations around the world provide records of waves that have traveled through the Earth along many different paths. By examining these records, it is possible to determine how the properties of rocks change with depth below the surface. The following characteristics of seismic waves are particularly useful in studying the inside of the Earth.

1. Seismic waves travel outward in all directions from the focus of an earthquake. By examining what happens to the waves as they pass through the Earth—how they slow down, speed up, and refract (bend)—scientists get a picture of the Earth's interior.
2. P waves generally travel faster than S waves, but the exact velocity of a seismic wave depends on the physical properties of the material it is traveling through. The velocities of seismic waves in different materials can be determined by laboratory experiments. Seismic waves travel at greater velocities through more compact materials. For this reason, seismic wave paths through the Earth are curved; the waves travel faster and faster with greater depth, because the materials at depth are more compact. Seismic waves also travel more quickly through cold materials than through hot materials.
3. P waves can be transmitted through solids and liquids, but S waves cannot be transmitted through liquids.

Imagine what happens when a seismic wave traveling through a homogeneous material encounters a sharp boundary with a material that has different physical properties. Let's say the boundary separates a rock unit of high seismic velocity from a rock unit of low seismic velocity. As it crosses the boundary the wave will slow down, causing its path to bend (Figure 4.10). This bending is called **refraction.** The same process causes light waves to bend when they pass from air to water.

You can demonstrate refraction of light waves by half-filling a glass with water. Place a pencil in the glass so that half of the pencil is submerged in the water. Look at the pencil from the side; does it look bent? The light rays traveling from the pencil through the water are bent—that is, refracted—when they pass from the water into the air.

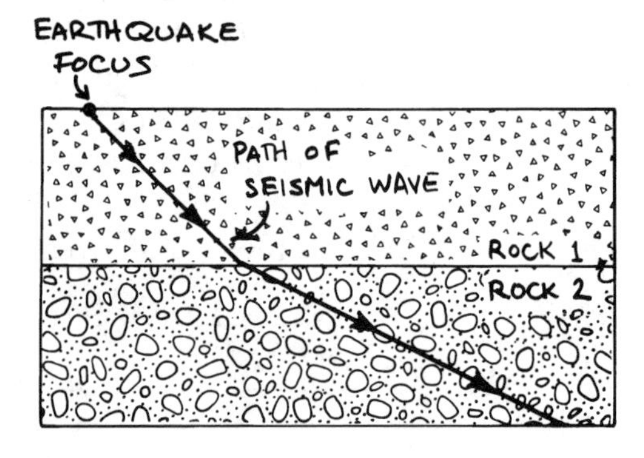

Figure 4.10

When P waves encounter the boundary between the mantle and the core, their paths are refracted quite sharply (Figure 4.11). The properties of the core are very different from those of the mantle, and P waves begin to travel much more slowly as soon as they hit the core. This causes a **P-wave shadow zone,** an area on the Earth's surface opposite the earthquake's focus, where P waves do not arrive as expected because they have been slowed down by their passage through the core.

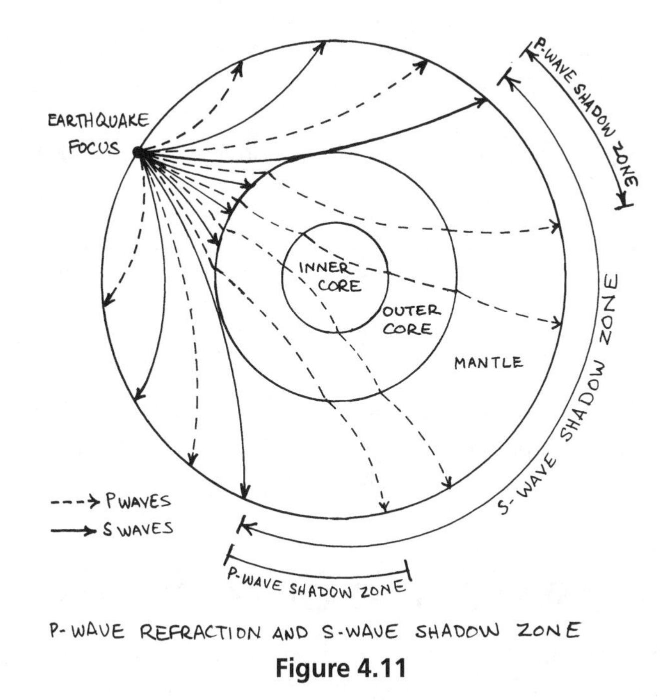

P-WAVE REFRACTION AND S-WAVE SHADOW ZONE

Figure 4.11

When S waves reach the core-mantle boundary, they are not refracted; they are blocked altogether. This creates a large **S-wave shadow zone** on the surface of the Earth opposite the earthquake's epicenter. Recall that S waves cannot be transmitted through liquids. From this we can deduce that the outer core of the Earth must be a liquid, which blocks the passage of S waves through the center of the Earth.

As more sophisticated seismic equipment was developed, scientists began to discover more boundaries and layers within the Earth. Today, seismologists use seismic waves to probe the Earth's interior in much the same way that doctors use X rays and computerized axial tomography (CAT) scans. In CAT scanning, a series of X rays along successive planes create a three-dimensional picture of the inside of the body. Similarly, seismic tomography allows seismologists to superimpose many two-dimensional seismic "snapshots" to create a three-dimensional image of the inside of the Earth.

We now know that there are many different layers inside the Earth, creating numerous boundaries where the velocities of seismic waves change abruptly. These are called **seismic discontinuities.** Some seismic discontinuities mark boundaries between layers of materials with different chemical compositions. Other discontinuities may represent boundaries between rocks of the same chemical composition but different physical properties. For example, the contrast in physical properties between a cold, hard, sinking lithospheric plate and the surrounding hot mantle could cause a discontinuity in seismic velocities. The transition from a low-pressure crystal structure to a denser, high-pressure crystal structure could also cause a seismic discontinuity. Seismic studies allow scientists to map the locations of seismic discontinuities, the distribution of hot and cold masses, and the distribution of dense and less dense materials inside the Earth.

Why are the paths of seismic body waves through the Earth curved instead of straight?

___

___

___

Answer: Body waves travel at greater and greater velocities with greater depth in the Earth, because the materials at depth are more compact. The steadily increasing velocity of the seismic waves causes the path to curve.

14 EARTHQUAKES AND PLATE BOUNDARIES

Earthquakes have helped scientists delineate the boundaries of lithospheric plates (Figure 4.12). As you know, most earthquake activity occurs along plate margins;

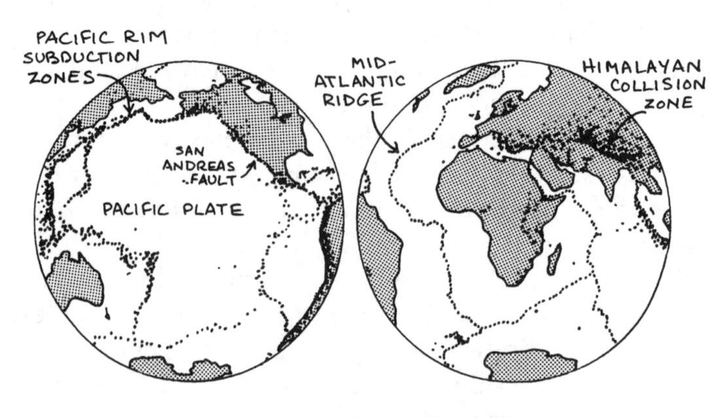

Figure 4.12

furthermore, different types of earthquakes occur at different types of plate margins. For example, earthquakes along divergent margins are usually fairly weak, with shallow foci. This is consistent with the presence of molten material just under the surface along divergent margins like midocean ridges, because earthquakes can occur only in rocks that are cold and brittle enough to break.

Other types of plate margins also have characteristic earthquake activity. Transform fault margins like the San Andreas Fault have earthquakes that occur at shallow to intermediate depths. These earthquakes can be very powerful, consistent with the fact that the blocks on either side of the fault are grinding past each other. Deep-focus earthquakes in collision zones, where continents crumple into great mountain ranges, also can be very powerful. In subduction zones, the entire surface along which one oceanic plate moves downward relative to the other plate is marked by powerful earthquakes (Figure 4.13). Earthquakes usually have shallow foci near the oceanic trench, where subduction begins. The earthquake foci become deeper and deeper along the descending edge of the subducting oceanic plate. These zones of shallow- to deep-focus earthquakes, called **Benioff zones,** first alerted scientists to the presence and geometry of subduction zones.

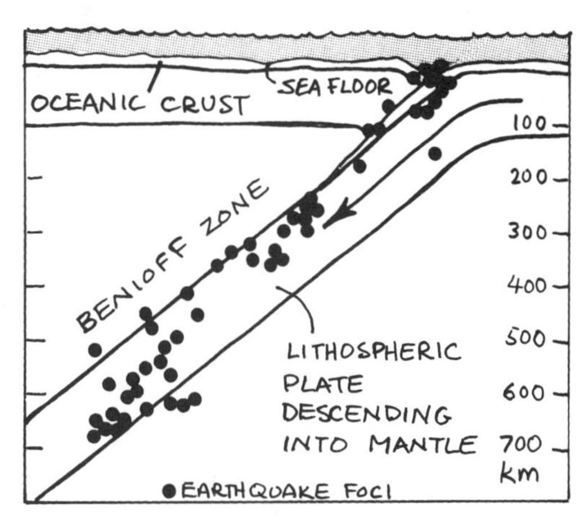

EARTHQUAKES IN SUBDUCTION ZONES
Figure 4.13

What kinds of earthquakes are characteristic of transform fault margins?

Answer: Shallow- to intermediate-focus earthquakes, which can be very powerful.

15 SUMMARY: LAYERS OF EARTH

Let's complete our study of the Earth's interior with a brief summary and review, starting with the crust and working down to the innermost layers (Figure 4.14). Keep in mind that some boundaries inside the Earth separate layers with different compositions, whereas others separate layers with the same composition but different physical properties.

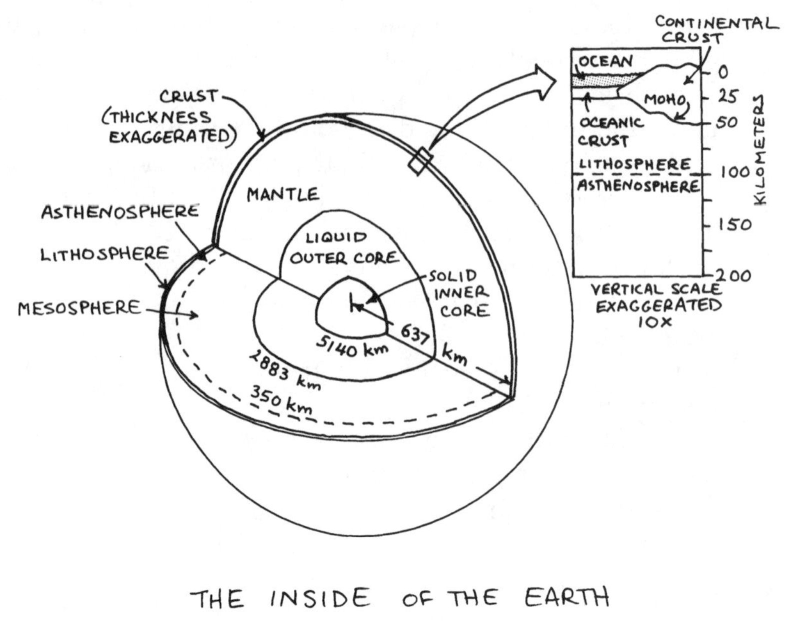

THE INSIDE OF THE EARTH

Figure 4.14

The outermost layer of the Earth is the crust. The thickness of the crust varies from an average of 8 km (5 mi) for oceanic crust to an average of 45 km (about 30 mi) for continental crust. Even at its thickest (about 70 km, or 44 mi), the crust is extremely thin compared to the Earth as a whole—it's like an eggshell, or the skin of an apple. About 95 percent of the crust is igneous rock, or metamorphic rock derived from igneous rock. In general, the rocks of the crust are less dense than the material that makes up the Earth's interior; this is the material that "floated" to the top during planetary differentiation. Continental crust is made mostly of granitic rocks, which are very low in density. Oceanic crust is made of basalt, which is denser than granitic rocks. The outermost layer of the crust is a thin veneer of sediment and sedimentary rock. Thus, although the crust is primarily igneous rock, most of what we actually see at the surface is sedimentary.

The boundary that separates the crust from the mantle was the first internal boundary to be discovered using seismic techniques. It was named the **Mohorovičić discontinuity** after the seismologist who discovered it in 1909, but we usually call it the **Moho.** The Moho is an example of a boundary between two layers of rock with differing compositions.

The mantle extends from the Moho to the core. About 80 percent of the volume of the Earth is contained in the mantle. Geologists believe that the mantle consists primarily of iron- and magnesium-silicate minerals. The upper part of the mantle has a composition similar to the rock peridotite, an igneous rock that consists primarily of the iron-magnesium silicate minerals olivine and pyroxene.

Extending from about 100 to 350 km (about 60 to more than 200 mi) below the surface is the asthenosphere. In this zone the rocks are near the temperatures at which melting begins, so they have little strength. Within the asthenosphere is a discontinuous zone in which P and S waves slow markedly. This is called the **low-velocity zone.** It is believed to result from the presence of melted rock. The composition of the asthenosphere appears to be the same as that of the mantle just above and below it. Thus, the asthenosphere is an example of a layer whose distinctiveness is based on its physical properties, not on its composition.

The lithosphere comprises the outermost 100 km (60 mi) of the Earth, which includes the crust and the part of the mantle just above the asthenosphere. The rocks of the lithosphere are cooler, more rigid, and much stronger than the rocks of the asthenosphere, which are weak and easily deformed, like butter or warm tar. In plate tectonic theory, it is the entire lithosphere (not just the crust) that forms the plates. The movement of lithospheric plates is facilitated by the presence of the underlying weaker rocks of the asthenosphere.

The rest of the mantle, from the bottom of the asthenosphere to the core-mantle boundary, is the **mesosphere** (from the Greek for "middle sphere"). Seismic discontinuities show that there is layering within the mesosphere. One important discontinuity occurs at 400 km (250 mi) and another at 670 km (415 mi) depth. These discontinuities reveal the existence of boundaries or transitions within the mantle, but the nature of the boundaries is not well understood. They may be compositional boundaries, that is, transitions from one type of rock to another. Or they may result from differences in physical properties. For example, when the mineral olivine is squeezed at a pressure equal to that found at a depth of 400 km, the atoms rearrange into a mineral structure that is more compact but still has the same chemical composition—a high-pressure polymorph of olivine. (Polymorphs are discussed in chapter 2.) It may be this change to a more compact crystal form that causes the 400-km seismic discontinuity.

It is important to remember that the mantle is basically solid rock, except for the small pockets of melt in the low-velocity zone. We know that the mantle must be solid because both P waves and S waves can travel through it. As you know from chapter 1, however, the mantle is convecting, even though it is solid. The pressures and temperatures deep within the Earth are so high that even solid rock can flow, although very, very slowly.

At a depth of 2,883 km (1,787 mi), another change in seismic velocity occurs. This is the core-mantle boundary, which represents a change in both composition and physical properties of the rocks. The core, the innermost of the Earth's compositional layers, is the densest part of the Earth. It consists of material that "sank" to the center during planetary differentiation. The Earth's core is composed primarily of iron-nickel metal. The S-wave shadow zone (and other evidence) tells us that the outer core, from 2,883 to 5,140 km (1,791 to 3,187 mi) in depth, must be liquid. Within this is the inner core, where the pressure is so great that iron is solid in spite of the very high temperature. The main difference between the inner and outer core is not one of composition; the compositions are believed to be virtually the same. The difference is in physical state: one is a solid, the other a liquid. As heat escapes from the core and works its way to the surface, the solid inner core must be crystallizing and growing larger, but very slowly.

Give one example of a boundary between layers of differing composition within the Earth, and one example of a boundary between layers of the same composition but differing physical properties.

Answer: Examples of boundaries between layers of differing composition include the crust-mantle boundary (the Moho) and the core-mantle boundary. Examples of boundaries between layers of differing physical properties include the lithosphere-asthenosphere boundary, the 400-km seismic discontinuity in the mantle, and the boundary between the inner and outer core.

Now you have learned some things about the interior of the Earth and its properties and how scientists investigate them. Heat escaping from the Earth's interior is the major driving force for plate tectonics, so what happens far below affects the surface of the Earth profoundly. See how well you have learned the material in this chapter by taking the Self-Test.

SELF-TEST

These questions are designed to help you assess how well you have learned the concepts presented in chapter 4. The answers are given at the end.

1. The Earth's asthenosphere is ___________.
   a. a hot, "weak" layer of rock
   b. just below the lithosphere
   c. basically the same composition as the mesosphere
   d. All of the above are true.
2. The energy released by a Richter magnitude 7.5 earthquake is approximately how many times greater than the energy released by a magnitude 5.5 earthquake?
   a. 30
   b. 60
   c. 100
   d. 900

3. The term "seismic gap" refers to ___________.
   a. a region where no seismic activity occurs
   b. a portion of a seismically active fault along which no large earthquakes have occurred recently
   c. a method for predicting seismic sea waves (tsunamis)
   d. the point of rupture of stressed rocks deep in the Earth (in the elastic rebound theory)

4. When a seismic wave crosses a boundary between layers of rock with differing physical properties, the seismic wave will bend. This bending is called ___________.

5. The ___________ scale for measuring earthquakes is based on eyewitness observations of earthquake vibrations and damage.

6. S waves travel more slowly than P waves through the Earth. (T or F)

7. The Moho is the boundary between the mantle and the core. (T or F)

8. A gravity anomaly is a location on the Earth's surface where the gravitational field of the Earth cannot be detected. (T or F)

9. What is the Earth's overall density? What is the average density of crustal rocks? What does this indicate about the distribution of material within the Earth?

____

____

____

10. How is gravity measured?

____

____

____

11. Why is drilling not a useful way for scientists to find out about the composition of the mantle?

____

____

____

12. How do diamonds reach the surface, and why are they useful in studies of the Earth's interior?

____

____

____

13. What are three secondary effects that are commonly associated with earthquakes?

____

____

14. Describe how triangulation is used to determine the location of the epicenter of an earthquake.

____

____

____

____

ANSWERS

1. d
2. d
3. b
4. refraction
5. Mercalli intensity
6. T
7. F
8. F
9. The overall density of the Earth is 5.5 g/cm^3. The average density of crustal rocks is 2.8 g/cm^3. This means that there must be a concentration of very dense material inside the Earth.
10. Gravity is measured with a gravimeter, which consists of a weight suspended from a spring attached to sensitive measuring devices. The stronger the pull of gravity, the more the weight is pulled down and the more it stretches the spring. The stretching of the spring provides a measure of the local pull of gravity.
11. The deepest hole ever drilled (almost 12 km [7.5 mi]) was in an area of thick continental crust, and didn't even come close to reaching the bottom of the crust. In areas where the crust is thinner, temperatures rise so quickly that drilling equipment would break down before reaching the bottom of the crust.
12. Diamonds are brought to the surface in xenoliths, carried by a type of magma (kimberlite) that originates deep within the mantle. By studying high-temperature, high-pressure minerals and magmas that form in the mantle, scientists can learn about the Earth's interior.
13. Any three of the following: aftershocks, landslides, liquefaction, fires, and tsunamis (technically, ground motion is a "primary" effect, not a secondary effect).
14. Three different seismic stations detect seismic waves from an earthquake. They can't tell from which direction the waves are coming, but they can calculate their distance

from the earthquake from the difference in arrival times of S and P waves. (S waves travel through the Earth more slowly than P waves; the farther the waves have to travel, the more the S waves will lag behind the P waves. The arrival time difference is therefore a measure of how far the waves have traveled, that is, the distance from the epicenter to the seismic station.) After calculating the distance, each station draws a circle around itself on a map, with radius equal to the distance from the epicenter. The point where the three circles overlap is the location of the epicenter.

KEY WORDS

aftershock
Benioff zone
body wave
compressional wave
dynamo hypothesis
elastic rebound theory
epicenter
fault
focus (plural: foci)
foreshock
geochemistry
geophysics
gravity anomaly
inertia
liquefaction
low-velocity zone
Mercalli intensity
mesosphere
Mohorovičić discontinuity (Moho)
P wave
P-wave shadow zone
paleoseismology
refraction
Richter magnitude
S wave
S-wave shadow zone
seismic discontinuity
seismic gap
seismic moment magnitude
seismic wave
seismogram
seismograph
seismologist
seismology
shear wave
surface wave
tsunami
xenolith

5 Volcanoes and Igneous Rocks

Fires that shook me once, but now to silent ashes fall'n away
Cold upon the dead volcano sleeps the gleam of dying day.

—Alfred, Lord Tennyson

A blast of burning sand pours out in whirling clouds.
Conspiring in their power, the rushing vapours
Carry up mountain blocks, black ash, and dazzling fire.

—Lucilius

OBJECTIVES

In this chapter you will learn

- how and why rocks melt;
- how cooling and crystallization influence the properties of igneous rocks;
- how plutons and plutonic rocks are formed;
- what causes volcanism and the hazards associated with volcanic eruptions.

1 WHY DO ROCKS MELT?

If a rock becomes very, very hot, it may reach its melting temperature, that is, the temperature at which it begins to change from a solid to a liquid. You can take a rock into a specially equipped laboratory, heat it up, and watch it turn into a liquid. The process is similar to heating ice to its melting temperature (0°C, or 32°F) and watching it turn into water, except that rocks melt at much higher temperatures than ice, typically 800°C (almost 1,500°F) or more. Also, for reasons that will be revealed later in the chapter, rocks typically melt over a range of temperatures, rather than at one specific temperature like ice.

The fact that rocks can melt, and that the resulting material is similar to the molten material that is erupted from volcanoes, was demonstrated almost 200 years ago by Sir James Hall, an experimentalist who worked with James Hutton. We continue to study the processes involved in rock melting because it helps us understand why different types of volcanoes behave the way they do. It also helps us interpret the geologic processes involved in the formation of the wide variety of igneous rocks that make up the crust and mantle of the Earth.

When was it first realized that rocks can melt, and who demonstrated this?

Answer: Almost 200 years ago; Sir James Hall.

2 THE GEOTHERMAL GRADIENT

The interior of the Earth is hot. If you go into a mine and measure rock temperatures, you will find that the deeper you go, the higher the temperature. The rate at which temperature increases with depth in the Earth is called the **geothermal gradient** (Figure 5.1). Temperature rises quite rapidly with depth under the continents, reaching 1,000°C (more than 1,800°F) at a depth of just over 100 km (60 mi). The temperature increase under the oceans is even more precipitous, reaching 1,000°C at an even shallower depth. In other words, the geothermal gradient is steeper under the oceans, and hotter temperatures are reached at a shallower depth than under the continents.

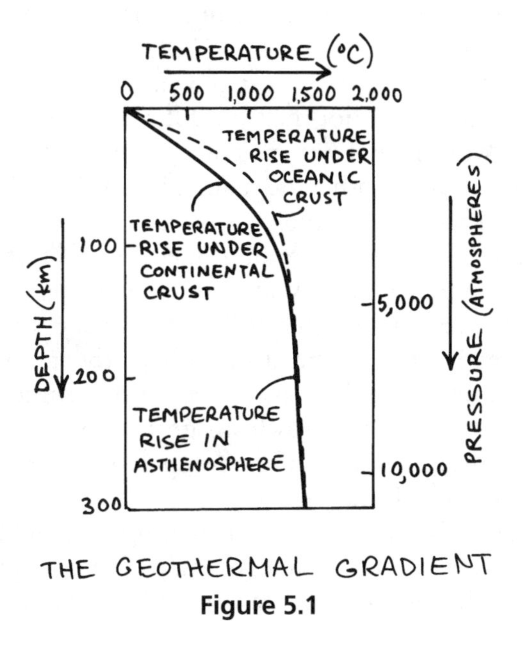

Figure 5.1

In chapter 4 you learned about the problems associated with drilling through oceanic crust. If you were to attempt to drill through both continental and oceanic crust to a depth of 50 km, what kinds of temperatures would you encounter? Use Figure 5.1 as a basis for comparison of continental and oceanic drilling.

Answer: As you can see from Figure 5.1, at a depth of 50 km under the continents you would encounter temperatures of about 700°C; the temperature at 50 km under the oceans would be close to 1,000°C.

3 TEMPERATURE, PRESSURE, AND WATER

The geothermal gradient brings up an interesting question. As noted, common rocks usually melt at temperatures above 800°C (1,500°F), depending on the exact composition of the rock. We know that temperatures in the Earth's mantle are well over 1,000°C (more than 1,800°F). So, why isn't the mantle entirely molten?

The answer is that pressure—which also increases with depth in the Earth—influences the temperature at which a rock will melt. As pressure increases, the melting temperatures of rocks and minerals also increase. For example, the mineral albite ($NaAlSi_3O_8$), a common rock-forming mineral and a member of the feldspar group, melts at 1,104°C (2,019°F) at the Earth's surface, where the pressure is low. However, at a depth of 100 km, where the pressure is 35,000 times greater, albite will not melt until it reaches 1,440°C (2,624°F). Whether a particular rock or mineral will melt at a given depth depends on the temperature *and* the pressure at that depth. The mantle is mostly solid because the pressures are so great that the rocks do not melt.

Here's another interesting question: if most minerals and rocks melt at *higher* temperatures when they are under pressure, do they melt at *lower* temperatures when the pressure is removed? The answer is yes. You can melt a mineral or a rock by increasing the temperature sufficiently at a given pressure; it is also possible to induce melting by decreasing the pressure sufficiently at a given temperature (Figure 5.2). If a rock moves from great depth toward the surface where pressure is lower it may melt, even if the temperature hasn't increased. This is called **decompression melting.**

The effect of pressure on melting is straightforward, provided the mineral is dry. However, when water (or water vapor) is present in a rock or mineral, it will melt at a lower temperature than a dry rock or mineral of the same composition.

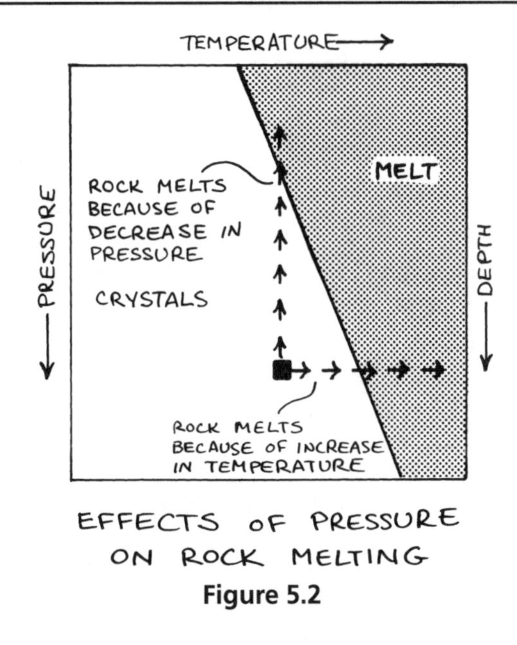

Figure 5.2

This is similar to the effect of salt on ice. If you live in a cold climate, you know that salt can melt the ice on an icy road. A mixture of salt and ice has a lower melting temperature than pure ice; try it! In the same way, a mineral-and-water mixture has a lower melting temperature than the pure mineral.

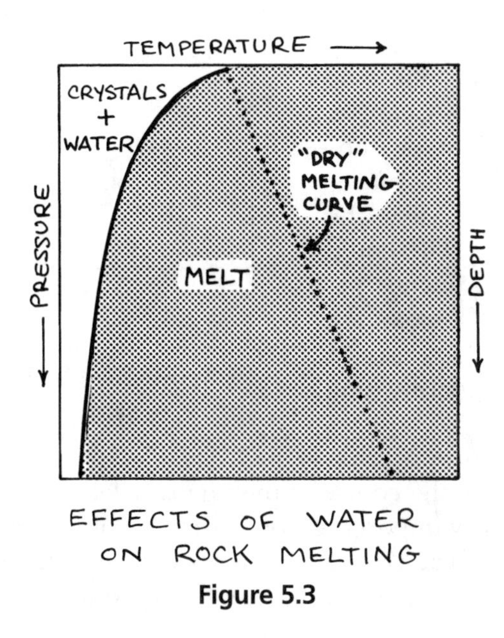

Figure 5.3

The effect of water on the melting temperatures of most rocks and minerals is opposite to the effect of pressure (Figure 5.3). In general, melting temperatures *increase* with increasing pressure but *decrease* with increasing concentration of water in the rock. What's more, the effect of water becomes more pronounced as pressure increases. In other words, as you increase the pressure on a wet rock or mineral, it will melt at lower and lower temperatures—exactly the opposite of what happens with a dry rock or mineral. This can be important in environments such as subduction zones, where water is carried down into the mantle by subducting oceanic plates.

What is the effect of pressure on the melting temperatures of most (dry) rocks and minerals?

____

____

Answer: Increasing pressure generally increases the melting temperatures of most (dry) rocks and minerals.

4 PARTIAL MELTING

Most rocks are composed of several different minerals. For this reason, rocks typically do not melt at one particular temperature the way a single mineral does. Instead, most rocks melt over a range of temperatures (Figure 5.4). For most common rocks, the temperature at which melting begins is lower than the melting temperatures of its constituent minerals. In other words, the melting temperature of the mixture (the rock) is lower than the melting temperatures of the individual ingredients or components of the mixture (the minerals). It's like the ice-and-salt mixture; the mixture begins to melt at a lower temperature than the melting temperature of either pure ice or pure salt.

As a rock melts across a range of temperatures, it will consist of some melted and some unmelted material; this is a **partial melt.** Eventually, when the temperature gets to be high enough, the entire rock will melt. But what if something interrupts this melting scenario while the rock is only partially molten? For example, the partially molten rock might be subjected to compression caused by the movement of lithospheric plates, so that the liquid portion is squeezed out. The liquid would consist mostly of components that melted at a low temperature, whereas the solid material left behind would consist of different components, with higher melting temperatures.

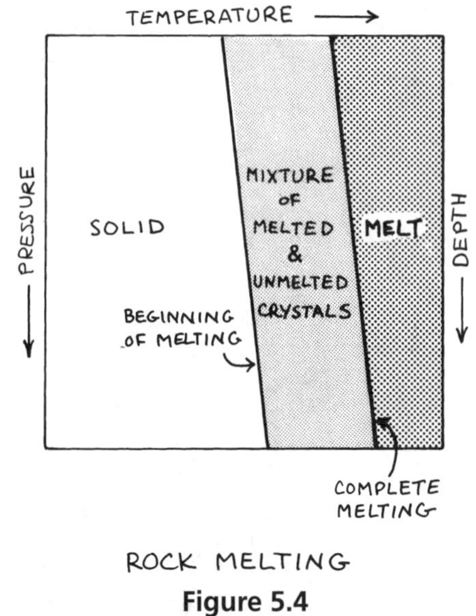

Figure 5.4

The process whereby a melt is separated from solid material during the course of melting is called **fractionation** (or "fractional melting"). It is a very important process, because it serves to separate a single rock into two batches of material—a melt and a solid—with entirely different compositions from each other (and, in fact, different compositions from the starting rock). Fractional melting contributes to the great diversity of rock types found in the Earth's crust.

What is a partial melt?

Answer: A rock that contains some melted material and some solid material.

5 MAGMA AND LAVA

Molten rock underground, usually with some floating mineral grains and dissolved gas, is called **magma.** Magma is typically less dense than the solid rock that surrounds it; therefore, it is buoyant and tends to rise toward the Earth's surface. If the magma makes it all the way to the surface, it is called **lava.** We can't study magma directly, but by studying lava we can make three important observations concerning molten rock.

1. It has a range of compositions in which silica (SiO_2) is usually dominant.
2. It is characterized by high temperatures.
3. It has the properties of a liquid, including the ability to flow.

Let's examine each of these observations in greater detail.

The chemical composition of magma is dominated by the most abundant elements—silicon (Si), aluminum (Al), iron (Fe), calcium (Ca), magnesium (Mg), sodium (Na), potassium (K), hydrogen (H), and oxygen (O). As you learned in chapter 2, O^{2-} is the most abundant anion, so we express the composition of a magma in terms of oxides, such as SiO_2, Al_2O_3, CaO, and H_2O. The most abundant component of common magmas is silica, SiO_2, which usually accounts for 45 to 75 percent of the magma by weight. Magmas in which silica is dominant are called **silicate magmas.**

The three most common types of silicate magma are **basaltic magma** (low silica, approximately 45 to 50 percent SiO_2 by weight), **andesitic magma** (intermediate silica, approximately 60 percent SiO_2 by weight), and **rhyolitic magma** (high silica, approximately 70 to 75 percent SiO_2 by weight). Of these, basaltic magma is by far the most common. Small amounts of dissolved gases (usually 0.2 to 3.0 percent by weight) are also found in most magmas. The main gases are water vapor and carbon dioxide. Despite their low abundances, these chemically active gases can strongly influence the properties of a magma.

The temperatures at which rocks melt are very high. Sometimes the temperature of a magma or a lava is even higher than the temperature at which the rock became completely molten. During eruptions of volcanoes such as Kilauea in

Hawaii and Mount Vesuvius in Italy, magma temperatures have been recorded from 800 to 1,200°C (1,472 to 2,192°F). Experiments using synthetic magmas in the laboratory suggest that under some conditions magma temperatures may be as high as 1,400°C (2,552°F).

Magmas are liquids, although some are very stiff and therefore flow very slowly. The property by which a substance offers resistance to flow is **viscosity;** the more viscous a substance, the less fluid it is. Water is an example of a very runny, low-viscosity liquid. Tar is an example of a high-viscosity liquid that does not flow easily. Some magmas are very fluid. Lava moving down a steep slope on Mauna Loa in Hawaii has been clocked at 64 km/h (40 mi/h), indicating a very low viscosity. However, some lavas are so viscous that they have trouble flowing at all.

The viscosity of a magma or a lava depends partly on its composition. In magmas, as in minerals, the SiO_4^{4-} anions link together, or polymerize (see chapter 2). Unlike the silicate anions in minerals, however, those in magmas form irregularly shaped groupings. As the groupings become larger, the magma becomes more viscous—that is, more resistant to flow—and behaves increasingly like a solid. In general, the higher the silica content of the magma, the larger the polymerized groups and the more viscous the magma. Temperature and gas content also affect the viscosity of magmas. The higher the temperature and gas content, the lower the viscosity and the more readily the magma flows. A very hot magma erupted from a volcano may flow easily, but as it cools it becomes more and more viscous and eventually stops flowing.

How does the silica content of a magma influence its viscosity?

___

___

Answer: The higher the silica content of a magma, the higher its viscosity. Silica facilitates the formation of polymerized groupings, which make it difficult for the magma to flow.

6 COOLING AND CRYSTALLIZATION: IGNEOUS ROCKS

When a magma cools and mineral grains start to form, the process is called **crystallization.** Just as melting influences the properties of magmas, cooling and crystallization influence the properties of the resulting igneous rocks. How fast a magma cools controls whether crystals will have a chance to grow, and how large they will be.

Crystals can grow only if their chemical constituents are able to move through the magma to the site where crystal growth is taking place. This is called **diffusion.** If a magma cools quickly, its rapidly increasing viscosity will prevent chem-

LAVA FLOWS. The two types of volcanic rocks in this photograph have the same composition (basalt), and they both formed from lava flows, but they have very different textures. The upper flow, which is rough, blocky, and fragmented, is a type of volcanic rock called aa (pronounced "AH-ah"). The lower flow, which has a smooth, ropy surface, is called pahoehoe (pronounced "pa-HOY-hoy"). The difference in texture is caused by a difference in the viscosities of the two lava flows. Aa forms from a cooler and more viscous lava, while pahoehoe forms from a hotter, more fluid lava. (Courtesy Brian Skinner)

ical elements from diffusing through the magma to the site of crystal growth. Therefore, the crystals will not grow very large. Thus, the rate of cooling and crystallization is very important in determining the **texture** of a rock—the size, shape, and arrangement of mineral grains that give the rock its overall appearance. Rate of cooling is the primary factor that distinguishes the two main classes of igneous rocks from one another: volcanic rocks cool and solidify quickly at or near the Earth's surface, whereas plutonic rocks cool and crystallize more slowly, deep underground.

The common volcanic and plutonic rocks are derived from silicate magmas that range from 45 to 75 weight percent silica. The chemical composition of a magma determines the types and amounts of minerals that will crystallize—that is, the **mineral assemblage.** Most common igneous rocks are dominated by one or more of the abundant rock-forming minerals: feldspar, mica, amphibole, pyroxene, and olivine. Quartz, feldspar, and some micas are light-colored minerals; the others are dark-colored. Rocks with large amounts of light-colored minerals are usually light in color and rich in silica; they are called **felsic** (from "feldspar" and "silica"). Rocks with large amounts of dark-colored minerals are usually dark in color and low in silica; they are called **mafic** (from "magnesium" and "ferric," or iron-rich; common chemical constituents of dark minerals). You may have heard low-silica rocks and magmas referred to as "basic" and high-silica rocks and magmas referred to as "acidic." These old-fashioned terms no longer reflect our understanding of the chemistry of rocks and magmas.

What is diffusion?

Answer: The process whereby chemical constituents move through a magma to the site where crystal growth is taking place.

7 FRACTIONAL CRYSTALLIZATION

Hundreds of different kinds of igneous rocks are found on the Earth. Most are rare, but the fact that they exist at all brings up an important question: how can all these different rock types form from a limited range of common magma compositions? It turns out that a single magma can crystallize into many different kinds of igneous rock, through **magmatic differentiation.**

When fractionation occurs during melting, the liquid portion of a partial melt is squeezed out, leaving behind a solid residue with a different composition. The same thing can happen during crystallization, in reverse. Crystallization occurs over a range of temperatures (the same as the temperature range over which melting occurs, but in reverse). Several different minerals will grow in a crystallizing magma, and those minerals will start to crystallize at different temperatures. Sometimes the newly formed crystals become separated from the melt (Figure 5.5).

How does this happen? If a magma is squeezed through a small opening, such as a fracture, crystals may clog the opening so that only the liquid can pass through; this is called "filter pressing." Crystal-melt separation can also happen when dense (heavy) crystals sink to the bottom of the magma, forming a solid layer of crystals covered by a melt of different composition; this is called "crystal settling." Crystals that are less dense (lighter) than the melt may float to the top; this is called "crystal flotation." Sometimes a newly forming crystal grows a protective rim around its outer edges; this is called "zonation" or "mantling," and it separates the core of the crystal from any further contact with the melt.

No matter how it happens, when crystals and liquids are separated from one another during crystallization it is called **fractional crystallization.** Through

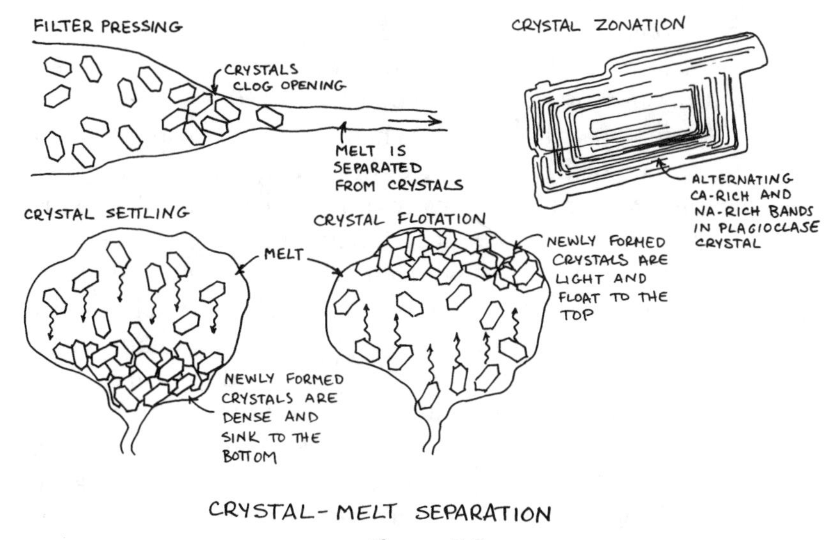

Figure 5.5

fractional crystallization, a single magma can be separated into a liquid and a solid with different compositions (and different compositions from the original magma, just as fractional melting creates batches of material with compositions that differ from each other and from the original rock). Different combinations of magma composition, partial melting, cooling rate, and fractional crystallization can lead to many, many variations in the resulting rock types. This is why we find so many different types of igneous rocks on the Earth.

What is magmatic differentiation?

Answer: The formation of many different rock types from a single magma composition.

8 BOWEN'S REACTION SERIES

A Canadian-born scientist, N. L. Bowen, first established the importance of magmatic differentiation by fractional crystallization. Because basaltic magma is by far the most common type of magma, he suggested that all other magmas may be derived from basaltic magma through fractional crystallization. He argued that in theory, a single magma could crystallize into a whole series of rock compositions, through fractional crystallization.

Bowen's experiments showed that the composition of the first plagioclase feldspar to crystallize from a basaltic magma is anorthitic (calcium-rich). But the composition of the plagioclase changes, becoming more albitic (sodium-rich) as crystallization proceeds. Bowen referred to this change as a "continuous reaction series," because the composition of the mineral in the cooling magma changes continuously by reacting chemically with the magma.

Here's how and why this chemical change happens. When a calcium-rich plagioclase crystallizes from a basaltic magma, it uses up some calcium from the magma, leaving the magma slightly depleted in calcium. The next crystal of plagioclase that crystallizes, therefore, will be slightly less calcium-rich than the first (and the temperature at which it crystallizes will be slightly lower as well, because the magma is cooling). The next plagioclase crystal, in turn, will be even less calcium-rich (more sodium-rich), and so on. In a continuous reaction series, *all* of the plagioclase crystals—even those formed early in the crystallization process—continually change and adjust their compositions to become more sodium-rich as the magma cools. They do this to stay in chemical equilibrium with the melt.

Through his experiments, Bowen identified several sequences of compositional changes involving other minerals besides plagioclase feldspars (Figure 5.6). One of the earliest minerals to form in a cooling basaltic magma is olivine. Olivine contains about 40 percent SiO_2 (silica) by weight, whereas basaltic magma contains about 50 weight percent SiO_2. When the silica-poor olivine grain crystallizes, it leaves the remaining magma just a little richer in silica. Eventually, as crystallization proceeds, the olivine is no longer in chemical equilibrium with the now more silica-rich melt. The olivine then reacts chemically with the melt to form a more silica-rich mineral, pyroxene. The pyroxene, in turn, eventually reacts with the melt to form amphibole, and the amphibole reacts with the melt to form an even more silica-rich mineral, biotite. A series of chemical reactions like this, in which early formed minerals form entirely new minerals through reaction with the melt, is called a "discontinuous reaction series."

If the reaction series is interrupted and the crystals become separated from the melt, fractional crystallization and magmatic differentiation occur. If the process of crystal-melt separation happens repeatedly, the composition of the residual melt

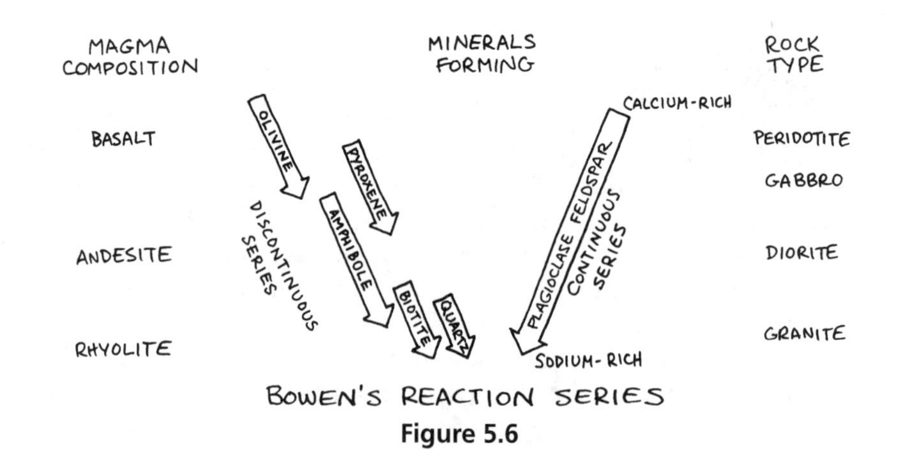

Figure 5.6

will become increasingly silica-rich and sodium-rich. Eventually, in theory, the final small fraction of melt will be very silica- and sodium-rich, with the composition of a rhyolitic magma.

Together, the continuous and discontinuous reactions identified by Bowen through his experiments are known as **Bowen's reaction series.** We now know that fractional crystallization rarely goes to these extremes. Therefore, the answer to the question that Bowen originally investigated—whether large volumes of rhyolitic magma can be formed from basaltic magma by fractional crystallization—is no. But careful study of almost any igneous rock will reveal that fractional crystallization has occurred at some point in its formation. Magmatic differentiation, fractional crystallization, and Bowen's reaction series play crucial roles in the formation of the great variety of igneous rocks we find on this planet.

What is the difference between a continuous reaction series and a discontinuous reaction series?

____

____

____

Answer: In a continuous reaction series, all crystals continually change and adjust their compositions as the magma cools, to stay in chemical equilibrium with the melt. In a discontinuous reaction series, early formed minerals react with the melt to form new minerals.

So far we have examined melting and its influence on the properties of magmas, and crystallization and its influence on the properties of igneous rocks. Let's turn to some of the processes involved in the formation of specific plutonic and volcanic rocks.

9 PLUTONIC ROCKS

Plutonic rocks (after Pluto, Greek god of the underworld) are also called "intrusive" rocks, because the magmas they came from intruded into the surrounding rocks. The minerals in plutonic rocks cool and crystallize slowly under the ground, so they usually have time to form relatively large, recognizable crystals. Thus, the texture of plutonic rocks is typically coarse-grained, or **phaneritic,** and individual mineral grains can be seen without a microscope. A plutonic rock with unusually large mineral grains (larger than 2 cm, or about 1 in) is called a **pegmatite.** Some mineral grains in pegmatites are huge—several meters or more in length. Pegmatites typically form during the last stage of crystallization of a plutonic rock body when there is a buildup of gas, including water vapor, in the remaining magma. The vapor facilitates the growth of large crystals by allowing chemical components to diffuse very quickly to the growing crystal surfaces.

The low-silica end of the compositional spectrum of plutonic rocks is represented by **gabbro,** a relatively coarse-grained, dark-colored rock with large amounts of pyroxene, plagioclase feldspar, and olivine. Gabbro has the same composition as the volcanic rock basalt, and it crystallizes from magma of basaltic composition. **Diorite** is an intermediate-silica plutonic rock. **Granite** and **granodiorite** represent the felsic, high-silica end of the compositional spectrum. They differ from each other mainly in the type of feldspars they contain: granites contain primarily alkali feldspar, whereas plagioclase is the dominant feldspar in granodiorites. Granite and granodiorite are common rocks in the continental crust, especially in the cores of mountain ranges.

What is a pegmatite?

Answer: An unusually coarse-grained plutonic rock.

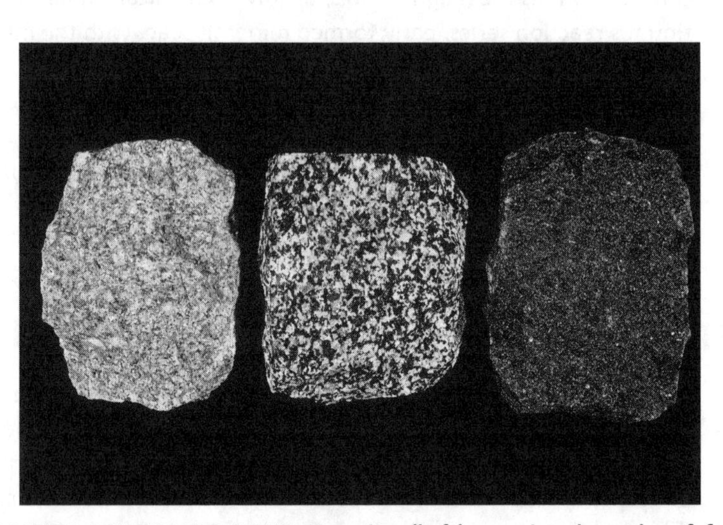

COARSE-GRAINED IGNEOUS ROCKS. Granite (left) consists largely of feldspar and quartz and is light in color (felsic) and rich in silica. Gabbro (right) consists of olivine, pyroxene, and calcium-rich feldspar, and is dark in color (mafic) and low in silica. Diorite (center) is intermediate in both color and silica content; it consists of feldspar and pyroxene or amphibole but does not contain either quartz or olivine. (Courtesy Brian Skinner)

10 PLUTONS AND PLUTONISM

All bodies of intrusive igneous rock, regardless of shape or size, are called **plutons** (Figure 5.7). The magma that forms a pluton did not originate at the location where we find the pluton. Rather, it intruded into the surrounding rock, which is called the "country rock."

A **batholith** (from the Greek words meaning "deep rock") is the largest kind of pluton. It is a large, irregular body of igneous rock that cuts across the layering of the country rock into which it intrudes. The largest batholith in North America is the Coast Range Batholith of British Columbia and Washington, which is about 1,500 km (900 mi) long. **Stocks** are also irregularly shaped plutons, but their maximum dimension is 10 km (about 6 mi). Some stocks are the tops of partly eroded batholiths. Most stocks and batholiths are granitic or granodioritic in composition. Batholiths are probably formed by extensive partial melting of the lower continental crust. The huge magma bodies that form batholiths move upward by melting or pushing overlying rocks out of the way.

A **dike** is a sheetlike body of igneous rock that cuts across the country rock layering. A dike forms when magma squeezes into a crosscutting fracture and solidifies. Tabular bodies of intrusive igneous rock that are parallel (or "conformable") to the layering of the country rock are called **sills.** Dikes and sills range from very small to very large. The Great Dike in Zimbabwe is a mass of gabbro nearly 500 km (more than 300 mi) long and 8 km (more than 5 mi) wide. The Palisades Sill is a large sill made of gabbro. It is about 300 m (almost 1,000 ft) thick, and forms the cliffs that line the Hudson River opposite New York City.

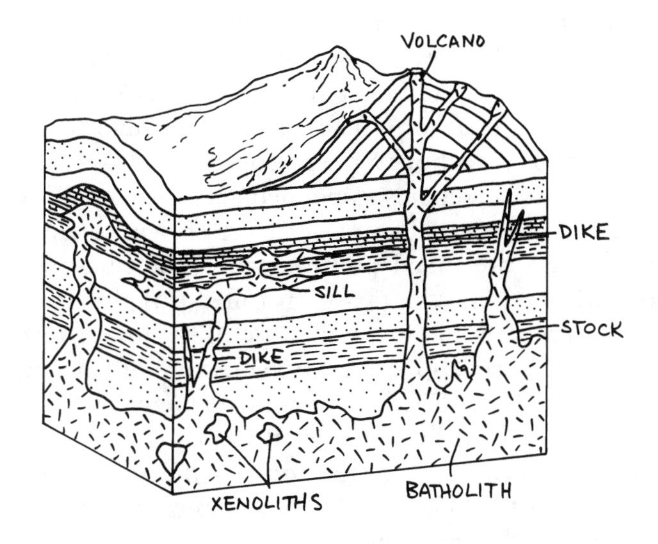

PLUTONS

Figure 5.7

There is a close link between plutonic rocks and volcanoes. Beneath every volcano is a complex network of channels and chambers through which magma reaches the surface. When a volcano stops erupting, the magma in the channels solidifies into plutonic rocks of various types. Some of these take the form of dikes or sills. **Volcanic pipes** are the remnants of channels that originally fed magma to the volcanic vent. A pipe that has been stripped of its surrounding rock by erosion is called a **volcanic neck.**

What is the difference between a dike and a sill?

___

___

Answer: A dike cuts across layering in the country rocks that it intrudes; a sill is intruded parallel to layering in the country rocks.

11 VOLCANIC ROCKS

When lavas and other volcanic materials cool and solidify, they become volcanic rocks. They are also called "extrusive" rocks because the magmas they formed from were extruded onto the Earth's surface. Lavas that erupt from volcanoes cool and solidify very quickly, because the temperature of the air (or water) is much lower than that of the lava. Sometimes lava cools so quickly that recognizable crystals do not form. The resulting noncrystalline rock is called **volcanic glass,** or "obsidian." Even when crystals do form in solidifying lavas, the crystals are typically very small—they simply don't have enough time to grow very big before the rock completely solidifies. Thus, volcanic rock is commonly very fine-grained, or **aphanitic** in texture. Usually, it is impossible to see the individual crystals in aphanitic rocks without a microscope.

If a magma contains suspended crystals that are extruded along with the lava, the resulting rock will consist of large crystals surrounded by aphanitic rock. The large crystals are called **phenocrysts.** The fine-grained part of the rock is the **groundmass,** and the texture is called **porphyritic.** Porphyritic texture is very common in volcanic rocks; it is useful because it reveals the history of the rock. Dissolved gas can also influence the texture of volcanic rock. As lava cools, its viscosity increases and it becomes increasingly difficult for gas bubbles to escape. When the lava finally solidifies, the last gas bubbles become trapped. This leaves holes, called **vesicles,** in the rock. If the magma is highly charged with gas, a frothy mass of bubbles may form, resulting in a spongy-textured, low-density volcanic rock called **pumice.**

Basalt is the dominant volcanic rock in oceanic crust. In fact, it is the most common volcanic rock type in the solar system, as far as we know. Basalt is a dark,

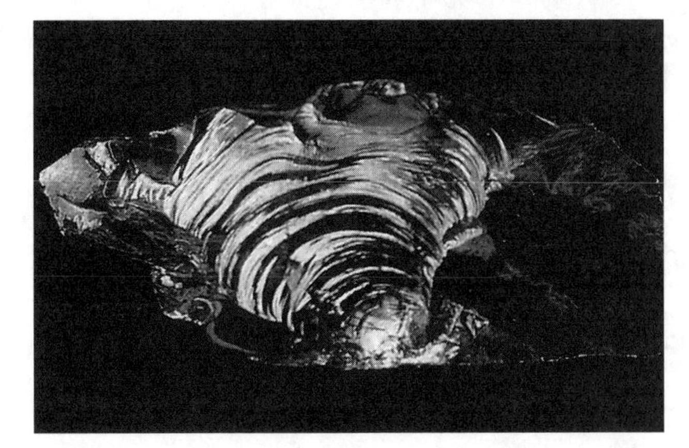

VOLCANIC GLASS. This sample of obsidian, a naturally formed volcanic glass, is from the Jemez Mountains in New Mexico. The curved ridges are typical of the fracture pattern in glass broken by a sharp blow. The specimen is 10 cm (4 in) across, and the glass is rhyolitic in composition. (Courtesy William Sacco)

fine-grained volcanic rock with low silica content. The mafic minerals olivine and pyroxene may make up more than 50 percent of the volume of the rock; the rest is usually plagioclase feldspar. **Andesite** is an intermediate-silica volcanic rock, with lots of feldspar and some amphibole or pyroxene. It is named for the Andes, the major volcanic mountain system of western South America. **Rhyolite** represents the felsic, high-silica end of the scale, consisting largely of quartz and feldspars. **Dacite** is similar to rhyolite but contains plagioclase feldspar, instead of the alkali feldspar found in rhyolite.

Each of the main volcanic rock types has a plutonic equivalent, as shown in Table 5.1. Gabbro and basalt are on the low-silica, mafic end of the scale; they crystallize from basaltic magmas. Andesite and diorite have intermediate silica contents; they crystallize from andesitic magmas. Rhyolite/dacite and granite/granodiorite are the most felsic and silica-rich; they crystallize from rhyolitic magmas.

Table 5.1 Plutonic and Volcanic Rock Equivalents

Silica Content	Melting Temperature	Viscosity of Magma	Color of Mineral Assemblage	Plutonic Rock	Volcanic Rock
High (≈ 70–75%)	Low (≈ 800°C)	High (not runny)	Felsic (light-colored)	Granite, granodiorite	Rhyolite, dacite
Intermediate (≈ 60%)	Medium (≈ 1,000°C)	Medium	Intermediate	Diorite	Andesite
Low (≈ 45–50%)	High (≈ 1,200°C)	Low (runny)	Mafic (dark-colored)	Gabbro	Basalt

On a separate sheet, make a table showing the common plutonic and volcanic rocks, their color (mafic or felsic), and their composition (approximate silica content).

ANSWER

Volcanic	Plutonic	Color	Composition
basalt	gabbro	mafic	low-silica
andesite	diorite	intermediate	intermediate-silica
rhyolite	granite	felsic	high-silica
dacite	granodiorite	felsic	high-silica

12 VOLCANOES AND VOLCANISM

A **volcano** (after Vulcan, the Roman god of fire) is a vent through which lava, rock debris, and gases are erupted. Volcanic eruptions are difficult to classify, partly because they change as they happen, either gradually (from month to month or year to year) or abruptly (from one day or one hour to the next). Some volcanoes erupt explosively; others are relatively quiet. The differences between nonexplosive and explosive eruptions are largely a function of the viscosity and dissolved gas content of the magma.

Recall that low-silica, basaltic magmas have relatively low viscosities; in other words, they are runny and tend to flow easily from volcanic vents. Successive basaltic lava flows build up to form broad, flat volcanic landforms with gently sloping sides, called **shield volcanoes.** An example is the Hawaiian volcano Mauna Kea. Although Mauna Kea rises only about 4.2 km (2.6 mi) above sea level, its height

SHIELD VOLCANO. Mauna Kea, a Hawaiian shield volcano 4.4 km (2.75 mi) in height, is seen here from neighboring volcano Mauna Loa. Note the gentle slopes formed by highly fluid basaltic lava flows. (Courtesy S. C. Porter)

from the seafloor is 10.2 km (more than 6.3 mi). This exceeds the height of Mount Everest (9.1 km, or 5.6 mi), making Mauna Kea the tallest mountain on Earth.

Some basaltic lava reaches the surface via elongate fractures called **fissures.** Low-viscosity lavas emerging from fissures on land tend to spread widely and may create vast, flat lava plains called **plateau basalts** or **flood basalts.** The Deccan Traps in India resulted from an extensive fissure eruption of this type. The Roza Flow, a great sheet of basalt in eastern Washington State, extends over an area of 22,000 square kilometers (km^2) (almost 8,500 square miles [mi^2]) and has a volume of 650 cubic kilometers (km^3) (156 cubic miles [mi^3]). Volcanic activity at seafloor spreading centers such as the Mid-Atlantic Ridge is often related to extensive fissuring.

If dissolved gas is present in a lava, it must escape somehow. Gas dissolved in a low-viscosity basaltic magma may cause the lava to bubble and fountain quite dramatically, especially at the beginning of an eruption. After the fountaining has died down, however, most basaltic eruptions are relatively quiet and are characterized by extensive lava flows. The higher the viscosity of the magma, however, the more difficult it is for dissolved gas to form bubbles and escape. As magmas move up toward the Earth's surface, the decrease in pressure allows the dissolved

VOLCANIC BOMB. This volcanic bomb was ejected during an eruption of a basaltic volcano at Craters of the Moon National Monument, Idaho. (U.S. Department of the Interior, National Park Service Photo)

gas to expand and form bubbles. In andesitic and rhyolitic magmas, more than in basaltic magmas, the gas bubbles are held back by the viscosity of the fluid. When the gas finally escapes, it usually does so explosively, blasting small bits of magma and other fragmental material in all directions.

Fragments ejected during an explosive volcanic eruption are called **pyroclasts** (from the Greek words *pyro,* "fire," and *klastos,* "broken") or **tephra** (from the Greek word for "ash"). Pyroclasts are classified on the basis of size: volcanic bombs are the coarsest, ranging up to the size of a small car, and volcanic ash is the finest type of tephra. Volcanic ash is not the same as the ash that forms from the burning of wood; rather, volcanic ash consists of particles of volcanic glass so small they may be distinguishable only under a microscope. Rocks that form by consolidation of pyroclasts are called **pyroclastic rocks.** Loose tephra particles can be converted into pyroclastic rock by cementing with another mineral or by the welding together of hot, glassy particles.

If a mixture of hot pyroclastic material and gas is denser than air, it may rush down the flanks of the volcano in a **pyroclastic flow,** also called a "glowing avalanche" or "nuée ardente." Pyroclastic flows are extremely dangerous because of their intense heat and the great speed at which they surge down the volcano's slopes. The most violent type of pyroclastic eruption is a **Plinian column,** a tur-

STRATOVOLCANO. Mount Mayon, an active stratovolcano in the Philippines, has steep slopes, formed largely of fragmental volcanic debris and viscous lava flows. (Courtesy of Solkoski/Terraphotographics/BPS)

bulent mixture of hot gas and tephra that rises explosively and buoyantly into the cooler air above the vent. Plinian eruption columns (named for the Pliny the Elder, who died during a major eruption of Mount Vesuvius in A.D. 79) may reach heights as great as 45 km (about 30 mi). Sometimes, instead of emerging from the top of the volcano, an explosive column may burst out from the side in a **lateral blast.** This happened in May 1980, during the modern eruption of Mount Saint Helens in Washington (Figure 5.8).

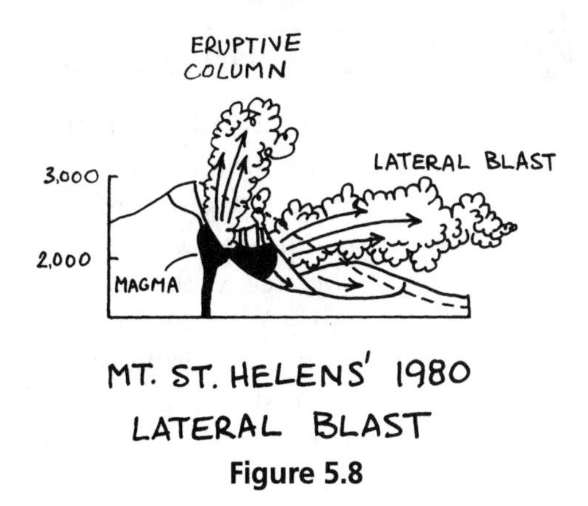

Figure 5.8

Explosive volcanoes tend to build volcanic landforms with steep sides. **Stratovolcanoes,** also called "composite" volcanoes, are composed of solidified lava flows interlayered with pyroclastic material. The steep, snow-capped peaks of classic stratovolcanoes like Mount Fuji in Japan and Mount Mayon in the Philippines are among the most beautiful sights on Earth. The majority of andesitic and rhyolitic volcanoes in continental and subduction zone settings are stratovolcanoes.

Other volcanic landforms and features can be associated with both explosive and nonexplosive eruptions. Near the summit of most volcanoes is a **crater,** a funnel-shaped depression from which gas, tephra, and lava are erupted. Some volcanoes have a much larger depression known as a **caldera,** a roughly circular, steep-walled basin that may be several kilometers or more in diameter. Calderas can form explosively as a result of the partial emptying of a magma chamber. The unsupported roof of the chamber may collapse under its own weight. Crater Lake in Oregon occupies a caldera that formed after a great eruption about 6,600 years ago of a volcano we now call Mount Mazama. Tephra deposits from that enormous eruption can still be seen in Crater Lake National Park and over a vast area of the northwestern United States and southwestern Canada.

A **fumarole** is a vent from which volcanic gas is emitted. The most common gas is water vapor, but other components, such as evil-smelling sulfur, may be present as well. Groundwater that comes into contact with hot volcanic rocks is heated, creating thermal springs, some of which have become famous health spas. A thermal spring with a natural system of plumbing that causes intermittent eruptions of water and steam is a **geyser,** from the Icelandic word *geysir,* "to gush."

What is the most explosive type of eruption?

Answer: A Plinian column.

13 VOLCANIC HAZARDS

Since A.D. 1800, there have been 19 volcanic eruptions in which 1,000 or more people have died. Most volcanic hazards cannot be controlled, but their impacts can be mitigated by effective prediction methods. Lava flows, pyroclastic activity, gas emissions, and volcanic seismicity are "primary" hazards—they are directly associated with the movement of magma and the eruptive products of the volcano. Other effects are "secondary" or "tertiary"; they happen as a secondary result of the eruption, and may have longer-term effects.

Most volcanoes produce at least some lava flows, but these typically cause more property damage than deaths or injuries. In Hawaii, lava flows have erupted from Kilauea almost continuously for more than a decade. Homes, cars, roads, and forests have been buried by lava flows or burned by the resulting fires, but no lives have been lost. It is sometimes possible to control or divert a flow by building retaining walls, or by chilling the front of the flow with water.

Unlike slowly moving lava flows, hot, rapidly moving pyroclastic flows and lateral blasts may overwhelm people before they can run away. The most destructive pyroclastic flow in this century (for loss of life) occurred on the island of Martinique in 1902. A glowing avalanche rushed down the flanks of Mount Pelée at a speed of more than 160 km/h (99 mi/h), killing 29,000 people. In A.D. 79, many of the citizens of Pompeii and Herculaneum were buried under hot pyroclastic material from an eruption of nearby Mount Vesuvius. Many of the victims were killed by poisonous volcanic gas, and their bodies were later buried by pyroclastic material. In 1986, at least 1,700 people and 3,000 cattle lost their lives when poisonous gas was emitted from a volcano at Lake Nyos, Cameroon.

The movement of magma toward the surface of the Earth causes rocks to fracture, resulting in swarms of earthquakes. The turbulent bubbling and boiling of magma under the ground can lead to a specific type of high-frequency seismicity known as volcanic tremor.

Many secondary and tertiary hazards are associated with volcanic eruptions, especially explosive eruptions. A large eruption, particularly in a coastal or marine setting, displaces the seafloor and can lead to the formation of a tsunami. This happened during the 1883 eruption of Krakatau, in Indonesia. More than 36,000 people were killed, most of them by the effects of the tsunami rather than the eruption itself.

Pyroclastic material can cause hazardous effects long after an eruption has ceased. Rain or meltwater from snow at the volcano's summit can mix with volcanic ash and start a deadly mudflow, referred to as a **lahar.** In 1985, a small eruption of Nevado del Ruíz in Colombia melted part of the icecap on the volcano's summit. Mudflows were formed when the meltwater mixed with volcanic ash. Massive lahars moved swiftly down river valleys on the flanks of the volcano, killing at least 23,000 people. A related phenomenon is a volcanic **debris avalanche,** in which many different types of materials—mud, blocks of pyroclastic material, trees, and so on—are mixed together. Much of the damage from the 1980 eruption of Mount Saint Helens was caused by a devastating debris avalanche.

Volcanic activity can change a landscape permanently. River channels may be blocked, resulting in flooding or permanent diversion of water flow. Mountainous terrains may be drastically altered; in the 1980 eruption of Mount Saint Helens, for example, the entire top and side of the mountain were blown away. New land can be formed, like the black sand beaches of Hawaii, which are made of dark pyroclastic material.

Volcanic eruptions also alter the chemistry of the atmosphere. In fact, the Earth's atmosphere and oceans originated from the outgassing of volatile material through volcanoes. The atmospheric effects of major eruptions can include salty, toxic, or acidic precipitation; spectacular sunsets; extended periods of darkness; and stratospheric ozone depletion. Global cooling can result from the blockage of solar radiation by fine pyroclastic material and volcanic aerosols. A famous example of this occurred after the 1815 eruption of Mount Tambora in Indonesia, which caused three days of total darkness as far as 500 km (300 mi) from the volcano. The following year was so cool that it was called "the year without a summer"; average global temperatures fell more than 1°C (almost 2°F) below normal, and there were widespread crop failures.

Not all impacts of volcanism are negative, and it is no accident that many people live near active volcanoes. Periodic volcanic eruptions replenish the mineral content of soils, ensuring continued fertility. Volcanism provides geothermal energy, and is linked with the formation of some types of mineral deposits. Volcanoes also provide some of the most magnificent scenery on this planet.

Summarize the beneficial effects of volcanism.

____

____

Answer: Replenishment and fertilization of soils with minerals; geothermal energy; formation of ore deposits; beautiful scenery.

14 PREDICTING VOLCANIC ERUPTIONS

Volcanic eruptions can't be prevented, but sometimes they can be predicted. The first step is to identify a volcano as active, dormant, or extinct. An active volcano is one that has erupted within recorded history; a dormant volcano has not erupted in recent history; and an extinct volcano shows no signs of activity and is deeply eroded. Mount Pinatubo in the Philippines had been dormant for several hundred years prior to its eruption in 1991, when it became active once again.

Another important step in prediction is to identify the volcano's past eruptive style. Mount Pinatubo is surrounded by thick deposits of pyroclastic material, a sign that the volcano erupted violently in the past and might do so again. The composition and tectonic setting of the volcano are also important. For example, andesitic and rhyolitic volcanoes in subduction zones are more likely to erupt explosively than are basaltic volcanoes and fissure eruptions.

Scientists monitor potentially dangerous volcanoes for signs of activity. Some of the warning signs are changes in the shape or elevation of the ground, such as bulging, swelling, or doming caused by inflation of the underlying magma chamber; changes in the temperature of crater lakes, well water, or hot springs; changes in heat output at the surface; changes in the amount or composition of gas emitted from the volcanic vent; and sudden increases in local seismic activity. Some monitoring, such as the monitoring of seismic activity, is done with instruments placed directly on the flanks of the volcano. Other monitoring—such as monitoring gas emissions or the temperature of the ground surface—can be done by means of remote sensing from satellites.

What are some of the warning signs of an imminent volcanic eruption?

___

___

___

Answer: Changes in shape or elevation of the ground, such as bulging, swelling, or doming; changes in the temperature of crater lakes, well water, or hot springs; changes in heat output at the surface; changes in the amount or composition of gas emitted from the volcanic vent; and sudden increases in local seismic activity.

You have learned a lot of new vocabulary terms and some difficult concepts in this chapter. Try the Self-Test to see if you have mastered this material.

SELF-TEST

These questions are designed to help you assess how well you have learned the concepts presented in chapter 5. The answers are given at the end.

1. The two common igneous rocks derived from basaltic magma are ___________.
   a. basalt and granite
   b. basalt and andesite
   c. basalt and gabbro
   d. granite and rhyolite

2. Under the oceans, the geothermal gradient is ___________ it is under the continents.
   a. steeper than
   b. not as steep as
   c. generally about as steep as

3. Some magmas are more viscous than others because of their ___________.
   a. compositions
   b. temperatures
   c. silica contents
   d. All of the above are true.
   e. None of the above is true.

4. A volcano made up primarily of lava flows, with broad, gently sloping sides is called a(n) ___________ volcano.

5. Magma temperatures measured during eruptions range from ___________ to ___________.

6. Another term for a volcanic mudflow is nuée ardente. (T or F)

7. Under certain circumstances, rocks can melt even if the temperature doesn't increase. (T or F)

8. How do temperature and gas content affect the viscosity of a lava?

____

____

9. What is the difference between felsic and mafic rocks?

____

____

____

10. What is the difference between aphanitic and phaneritic textures?

____

____

____

11. What is the most common effect of water on the melting temperature of a rock?

____

____

____

ANSWERS

1. c
2. a
3. d
4. shield
5. 800°C; 1,200°C
6. F
7. T
8. In general, the higher the temperature and the higher the gas content, the lower the viscosity of a lava of a given composition.
9. Felsic (from "feldspar" and "silica") rocks are generally light-colored and silica-rich, and contain an abundance of the light-colored minerals feldspar, quartz, and muscovite. Mafic rocks (from "magnesium" and "ferric") rocks are generally dark-colored and low in silica, and contain an abundance of dark, magnesium- and iron-rich minerals.
10. Rocks with aphanitic texture are fine-grained; it is difficult or impossible to see and identify the individual mineral grains without the aid of a microscope. Rocks with phaneritic texture are coarse-grained; individual mineral grains are clearly visible by eye.
11. In general, a rock will melt at a lower temperature in the presence of water than a dry rock of the same composition at the same pressure.

KEY WORDS

andesite
andesitic magma
aphanitic
basalt
basaltic magma
batholith
Bowen's reaction series
caldera
crater
crystallization
dacite
debris avalanche
decompression melting
diffusion
dike
diorite
felsic
fissure
fractional crystallization
fractionation
fumarole
gabbro
geothermal gradient
geyser
granite
granodiorite
groundmass
lahar
lateral blast
lava

mafic
magma
magmatic differentiation
mineral assemblage
partial melt
pegmatite
phaneritic
phenocryst
plateau (flood) basalt
Plinian column
pluton
porphyritic
pumice
pyroclast
pyroclastic flow
pyroclastic rock
rhyolite
rhyolitic magma
shield volcano
silicate magma
sill
stock
stratovolcano
tephra
texture
vesicle
viscosity
volcanic glass
volcanic neck
volcanic pipe
volcano

6 Weathering and Erosion

The heights of our land are thus leveled with the shores; our fertile plains are formed from the ruins of mountains.

—James Hutton

Nothing under heaven is softer or more yielding than water, but when it attacks things hard and resistant there is not one of them that can prevail.

—Lao-tzu

OBJECTIVES

In this chapter you will learn

- how mechanical, chemical, and biological processes cause rocks to disintegrate;
- how the agents of erosion transport broken-down particles of rock;
- how wind, water, and ice sculpt and modify the landscape;
- what kinds of landforms are formed when sediment is deposited by wind, water, and ice.

1 WHY ROCKS FALL APART

All rocks on and near the Earth's surface are susceptible to **weathering,** the chemical and physical breakdown of rock exposed to air, water, or biological activity. Weathering takes place in a zone that extends downward below the Earth's surface as far as air, water, and microscopic organisms can readily penetrate—tens to hundreds of meters. The result of weathering is regolith, the loose surficial blanket of broken rock and mineral fragments. The size of fragments in the regolith ranges from microscopic to many meters across, but all have been loosened from solid bedrock below.

Because regolith is loose, it can be readily moved around. Wind, water, and ice constantly modify the Earth's surface, cutting away material here, depositing material there, sculpting and creating the familiar landscapes around us in the process. We use the term **erosion** to describe the combined processes of weathering and the subsequent transport of regolith by natural agents.

What is the difference between weathering and erosion?

____

____

Answer: Weathering is the physical and chemical breakdown of rock. Erosion is the combination of weathering and the transport of weathered particles.

2 MECHANICAL WEATHERING

Mechanical weathering is the physical breakdown of rock with no change in mineralogy or chemistry. One of the most important mechanical weathering processes is the development of **joints,** rock fractures along which no observable movement has occurred. Rock buried deep beneath the surface is subject to enormous pressures from the weight of overlying rock. As erosion removes material from the surface, the weight is reduced and the buried rock responds by expanding upward. As the rock expands, it fractures and forms joints. Sometimes large, curved slabs of rock peel away from the surface of the rock; this is called **exfoliation.** Joints and other structural openings are the main passageways through which water, air, and small organisms can penetrate rocks; this promotes further weathering.

The freeze-thaw cycle of water is another very important cause of mechanical weathering. When water freezes to ice, its volume increases by about 9 percent. (Almost all other substances undergo a *decrease* in volume upon freezing.) When temperatures fluctuate around the freezing point, water in the ground will alternately freeze and thaw. The volume increase with each freeze causes stress, which can eventually shatter the rocks. This process, the freezing of ice in a confined opening within a rock, causing the rock to split apart, is known as **frost wedging.**

The growth of salt crystals within rock cavities or along grain boundaries can sometimes generate enough force to cause rocks to split. This process occurs mostly in deserts, where salts are precipitated from groundwater as a result of evaporation. Forest fires can also be an effective way to break rocks. Because rock is a poor conductor of heat, an intense fire heats only a thin outer shell of the rock, which expands and breaks away.

Sometimes mechanical weathering is caused by biological agents. For example, the roots of a growing tree can widen cracks and eventually wedge apart

JOINTS. These sedimentary rocks are broken by two sets of joints. The steeply sloping sedimentary beds, once horizontal, have been tilted by the forces that created the Rocky Mountains. The two sets of joints are perpendicular to the sedimentary layers. The location is in Alberta, Canada. (Courtesy Brian Skinner)

bedrock. In much the same way, roots disrupt stone walls, sidewalks, and even buildings. Large trees swaying in the wind can also cause fractures to widen. When the trees are blown over they can disrupt the rock still further. Although it is difficult to measure, the total amount of rock breakage caused by plants must be very large.

What is a joint?

Answer: A rock fracture along which no observable movement has occurred.

3 CHEMICAL WEATHERING

Chemical weathering is the decomposition of rocks and minerals as a result of chemical reactions that change them to new minerals that are stable at the Earth's surface. Although chemical weathering is distinct and different from mechanical weathering, the two processes almost always occur together, and their effects are sometimes difficult to separate.

FROST WEDGING. This photograph shows the effects of frost wedging in eastern Greenland. Frost wedging is an example of mechanical weathering, in which water penetrates joints, freezes, expands, and wedges rocks apart. (Courtesy A. L. Washburn)

Chemical weathering is primarily caused by water that is slightly acidic. As raindrops fall through the air they dissolve carbon dioxide and so become a weak solution of carbonic acid (H_2CO_3). When weakly acidified rainwater enters the regolith, it may dissolve additional carbon dioxide from decaying organic matter and become more strongly acidified. Another way that rainwater can become strongly acidified is by interacting with **anthropogenic** (human-generated) sulfur and nitrogen compounds released to the atmosphere, forming solutions of sulfuric acid (H_2SO_4) and nitric acid (HNO_3) known as **acid rain.**

Acidified water reacts with and changes minerals in three principal ways.

1. **Hydrolysis.** In the process of hydrolysis, hydrogen ions enter a crystal and displace ions such as potassium (K^+), sodium (Na^+), and magnesium (Mg^{2+}). The displaced ions pass into solution and flow away in groundwater. All acids, including carbonic acid and the acids from acid rain, form hydrogen ions (H^+) when they are in solution. Ions in solution are not fundamentally different from ions in minerals, except that ions in solution can move randomly about and cause chemical reactions. The acids of acid rain are stronger than carbonic acid, so they produce more hydrogen ions. Thus, weathering rates due to acid rain are much faster than natural weathering rates. Eventually, the ions released by hydrolysis reach the sea; they are a major source of sea salts.
2. **Solution** (or **dissolution**). Sometimes minerals are completely dissolved and removed by flowing water. The most soluble (easily dissolved) of the rock-forming minerals are the carbonate minerals calcite and dolomite. Soluble substances dissolved from rocks, together with ions displaced during hydrolysis, are present in all groundwater and surface water. Sometimes their concentrations are so high that the water has an unpleasant taste.
3. **Oxidation.** Iron (and some other elements) can be chemically altered by oxidation, a process in which ferrous iron (Fe^{2+}) is changed to ferric iron (Fe^{3+}). Ordinary rust is a form of chemical weathering by oxidation of ferrous iron. Iron is present in many rock-forming minerals. When such minerals are chemically weathered the iron is released and oxidized. Oxidized iron can eventually form hematite (Fe_2O_3), a brick-red-colored mineral that gives most tropical soils of the world their distinctive red color.

Like mechanical weathering, chemical weathering can be greatly enhanced by biological activity. For example, burrowing animals such as ants, termites, and rodents bring rock particles to the surface, where they are more fully exposed to chemical weathering. More than 100 years ago, Charles Darwin calculated that earthworms bring particles to the surface at a rate in excess of 2.5 kg/m^2 (10 tons/acre) per year. Biochemical processes such as the decomposition of organic matter are also very important in the formation of soils (as you will learn in chapter 7).

How does rainwater become acidified?

____

____

Answer: Rainwater becomes acidified by dissolving carbon dioxide from the air, dissolving carbon dioxide from decaying organic matter in the regolith, or interacting with anthropogenic sulfur and nitrogen compounds.

4 CONDITIONS THAT AFFECT WEATHERING

Many factors influence the rate of weathering and the susceptibility of a rock to weathering, as shown in Table 6.1. Because specific minerals react differently to weathering processes, rock type and mineralogy are clearly important. Quartz, one of the most common rock-forming minerals, is also one of the most resistant to chemical weathering. This is because quartz is not affected by hydrolysis, which affects almost every other silicate mineral. Rocks rich in quartz, such as granite, are therefore slow to be affected by chemical weathering.

Weathering is also strongly influenced by the structure of the rock, especially the presence and spacing of joints. Even a rock that consists entirely of quartz may break down rapidly if it contains closely spaced joints that are susceptible to frost wedging. High temperatures and abundant rainfall also promote weathering, by enhancing the rate of chemical reactions.

Chemical weathering is more intense and extends to greater depths in warm, wet tropical climates than in cold, dry Arctic climates. In moist tropical lands, the effects of chemical weathering can be detected to depths greater than 100 m

Table 6.1 Factors That Influence Weathering

	Rate of Weathering		
	Slow ⟶		**Fast**
Mineral resistance	High (e.g., quartz)	Intermediate (e.g., mica)	Low (e.g., calcite)
Frequency of joints	Few (meters apart)	Intermediate (≈ 0.5–1.0 m apart)	Many (centimeters apart)
Steepness of slope	Gentle	Moderate	Steep
Vegetation	Dense	Moderate	Sparse
Temperature	Cold (avg. ≈ 5°C)	Temperate (avg. ≈ 15°C)	Warm/Hot (avg. ≈ 25°C)
Rainfall	Low (>40 cm/yr)	Intermediate (40–130 cm/yr)	High (>130 cm/yr)
Burrowing animals	Rare	Frequent	Abundant

(more than 300 ft). By contrast, in cold, dry regions and on high mountain peaks, chemical weathering proceeds very slowly, mechanical weathering is dominant, and evidence of weathering disappears at shallow depths.

On a steep slope the solid products of weathering are quickly moved away, continually exposing fresh bedrock. Slopes that are cleared of vegetation also weather more rapidly than vegetated slopes. Removal of vegetation can happen naturally—through landslides, for example—or as a result of activities such as clear-cut logging, slash-and-burn agriculture, and grazing. This can lead to accelerated weathering and erosion of newly exposed regolith.

What are some of the ways in which human activities can influence weathering?

____

____

Answer: Examples in the text include acid rain and removal of vegetation by clear-cut logging, slash-and-burn agriculture, and grazing. Can you think of any others?

5 TRANSPORT AND DEPOSITION

Erosion, as defined at the beginning of this chapter, is a general term that encompasses weathering plus the subsequent transport of weathered particles. Loose particles may be transported by flowing air, water, or ice, or they may move downslope under the influence of gravity; the latter is called **mass wasting.**

The ability of a fluid (whether air, water, or flowing ice) to pick up and transport particles depends on the velocity and turbulence of the fluid, as well as its viscosity. When a fluid flows slowly, all the fluid particles travel in parallel streams; this is called **laminar flow** (Figure 6.1). With increasing velocity, the movement becomes erratic and complex, giving rise to the swirls and eddies that characterize **turbulent flow.** Laminar flow can transport some particles, but turbulent flow is needed to pick up the coarsest particles and move them along the ground.

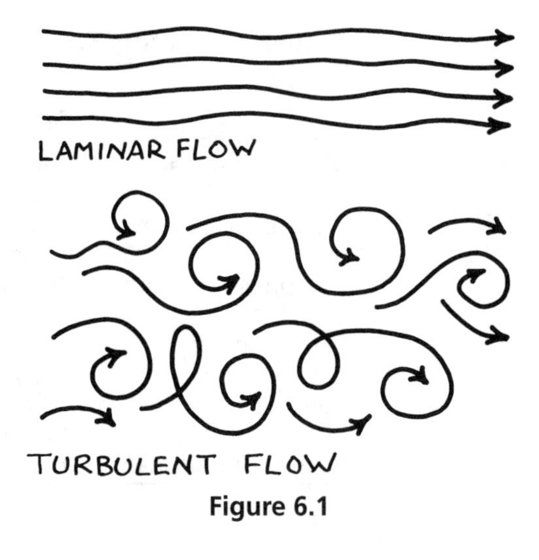

Figure 6.1

The velocity at which a flowing fluid becomes turbulent depends partly on its viscosity. The lower the viscosity the "runnier" the fluid, and the lower the velocity at which turbulent flow occurs. Air has a very low viscosity, and almost all airflow (wind) is turbulent. Water has a higher viscosity than air, but turbulent flow still predominates in streams. In turbulent flow, fine particles may be supported and carried indefinitely by the current, a mode of transport called **suspension** (Figure 6.2). The muddy character of many streams is caused by fine particles of silt and clay moving in suspension. As the velocity and turbulence of flowing air or water increase, coarser particles start to roll along the ground or stream bottom. Moving grains impact other grains, causing them to move forward in a series of short jumps along arc-shaped paths. This is called **saltation.**

The movement of flowing ice and the mechanisms through which ice carries particles are somewhat different from those in air or water, because ice—although it does flow—is a solid. The flow of glacial ice is basically laminar, but the high viscosity of ice allows it to carry much larger and heavier particles than either air or water.

When the velocity and turbulence of a flowing fluid decrease, the fluid loses its ability to carry particles, which will then settle out and accumulate as sediments. This is **deposition;** it happens when the wind velocity drops or when a swiftly flowing stream enters a still body of water such as a lake. The heaviest particles settle out first, while upward-flowing turbulent currents may still have sufficient velocity to keep the finer, lighter particles in suspension. In chapter 7, you will learn more about sediments and the depositional environments in which they accumulate.

What is the difference between laminar flow and turbulent flow?

Answer: In laminar flow, all the fluid particles move in parallel streams; turbulent flow is characterized by complex swirls and eddies.

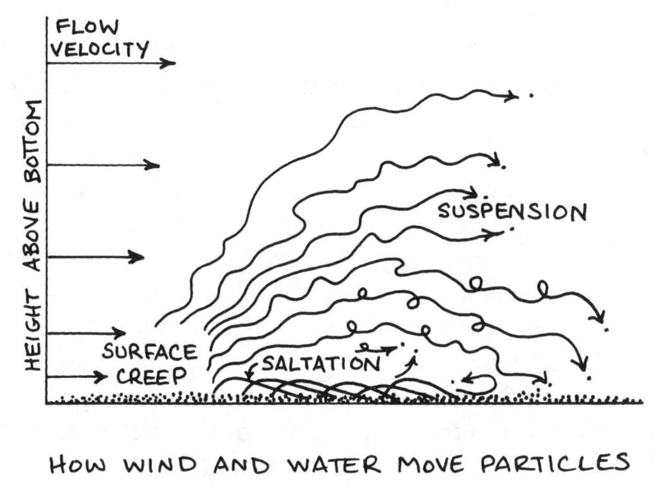

HOW WIND AND WATER MOVE PARTICLES

Figure 6.2

6 EROSION BY WATER ON THE LAND

Erosion by water begins as soon as rainfall hits the ground. It happens by impact as raindrops hit the ground and dislodge small particles, and by sheets of water moving overland during heavy rains. Overland flow on a slope quickly concentrates into channels, thus becoming a **stream,** a body of water flowing downslope along a clearly defined natural passageway. Streams carry most of the water that goes from the land to the sea, and they transport billions of tons of sediment to the ocean each year. They also carry the soluble salts released by chemical weathering of rocks, which play an essential role in maintaining the salinity of seawater.

A number of factors control stream behavior. The most important are the steepness (gradient) of the channel; cross-sectional area (width × depth) of the channel; velocity of water flow; **discharge,** the amount of water (width × depth × velocity) passing by a given point on the channel's bank during a period of time; roughness of the stream bottom (the "bed"); and **load,** the material being carried along by the stream. The load has three parts: **bed load,** particles that move along the bottom; **suspended load,** particles that are carried by suspension in the water; and **dissolved load,** dissolved substances that are a product of rock weathering.

All of these factors are interrelated. For example, if the gradient of a stream steepens, the velocity of flow is likely to increase as well. If the velocity is high, a greater load can be carried. If the discharge increases, the channel must handle more water; as a result, both the velocity of flow and the depth of the water in the stream will likely increase. In some cases, the width and depth of the channel may also increase. A flowing stream may increase the cross-sectional area of its channel by scouring the banks and the bottom. This scouring not only deepens and widens the channel, but also adds sediment to the load. When velocity eventually decreases, the sediment settles out, filling in the channel and allowing it to return to its original size.

The ability of streams to erode and carry particles of sediment is related to the way water moves through the channel. The higher the velocity and the more turbulent the flow, the greater the ability of the water to pick up particles and carry them away. When a stream loses velocity because of a change in gradient or a decrease in discharge, its transporting power drops and it deposits part of its load.

The shape of a channel reflects variations in both erosion and deposition; no two channels are exactly alike. The underlying rocks are important, too. For example, a stream may take a sudden bend or its gradient may increase when it passes from rock that is resistant to erosion into rock that is easily eroded. Straight channels are particularly rare. Even in a relatively straight channel, a line connecting the deepest parts of the channel typically does not follow a straight path, but wanders back and forth across the channel (Figure 6.3). The water velocity tends to be highest in the deepest parts of the stream. In places where the deepest water lies at one side of a channel, a deposit of sediment (a **point bar**) tends to accumulate on the opposite side because velocity is lower.

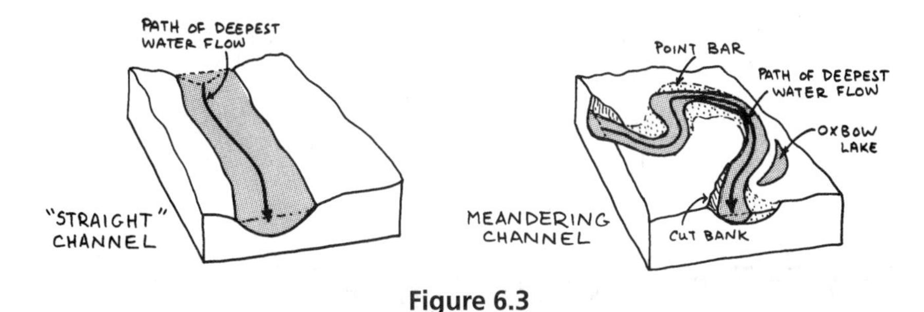

Figure 6.3

In many streams the channel forms a series of smooth bends called **meanders,** named after the Menderes River (in Latin, *Meander*) in southwestern Turkey, noted for its winding course. When the water in a meandering stream rounds a bend, the zone of highest velocity swings toward the outside of the channel. As water sweeps around the bend, turbulent flow causes undercutting and erosion of material where the fast-moving water meets the steep outer bank. Along the inner side of each meander, where the water is shallow and velocity is low, sediment accumulates to form a point bar. As a result, meanders slowly change shape and shift position along a valley as sediment is subtracted from and added to the banks.

MEANDERS. This aerial photograph shows well-developed meanders in the Laramie River, Wyoming. (Courtesy U.S. Geological Survey)

What are the three types of load in a stream?

Answer: Bed load, suspended load, and dissolved load.

7 EROSION BY WATER UNDER THE GROUND

Water can also cause erosion underneath the ground. As soon as rainwater infiltrates the ground to become **groundwater,** it begins to react with the minerals in the regolith and the bedrock, causing chemical weathering. Among the minerals of the Earth's crust, the carbonates are most readily dissolved. Carbonate rocks such as limestone are almost insoluble in pure water, but are easily dissolved by carbonic acid, a common constituent of rainwater. The attack occurs mainly along joints and other openings in the rock. When limestone weathers, it may be dissolved and carried away in slowly moving groundwater. In some carbonate terrains, the rate of dissolution is even faster than the average rate of erosion of surface materials by streams and mass wasting.

You can demonstrate the dissolution of a carbonate using an egg and some vinegar. Eggshells are partly composed of $CaCO_3$ (calcium carbonate), and vinegar is a weak acid. Submerge the egg in the vinegar and check it after about a day. You will find that the calcium carbonate in the shell has dissolved, leaving it soft and elastic.

When carbonate rock is dissolved by circulating groundwater, a cave may form. **Caves** are dissolution cavities that are closed to the surface, or have only a small opening. Cave formation begins with dissolution along interconnected fractures and bedding planes, where two different sedimentary rock units meet. A passage eventually develops along the most favorable flow route. The rate of cave formation is related to the rate of dissolution. As the passage grows and the flow of groundwater becomes more rapid and turbulent, the rate of dissolution also tends to increase. The development of a continuous passage by slowly moving groundwater may take up to 10,000 years, and enlargement of the passage to create a fully developed cave system may take as long as a million years. Terrains that are underlain by extensive cave systems are called **karst** terrains.

Sinkholes are dissolution cavities, like caves, but open to the sky. Some sinkholes are formed when the roof of a cave collapses. Others are formed at the surface, where rainwater is freshly charged with carbon dioxide and hence is most effective as a solvent. Some sinkholes form slowly; others form catastrophically. An example of the latter occurred in Winter Park, Florida, in 1972. In a period of just

10 hours, a sinkhole developed and consumed part of a house, six commercial buildings, several automobiles, and a municipal swimming pool. The total cost of the damage was over $2 million. Events as dramatic as the Winter Park sinkhole are rare, but sinkhole collapse is a common occurrence in areas underlain by carbonate rocks.

In what type of terrain is cave and karst formation most common?

Answer: A carbonate terrain.

8 EROSION BY WIND

Wind is an important agent of erosion, especially in arid and semiarid regions. Processes related to wind are called **eolian** processes after Aeolus, the Greek god of wind. Because the density of air is far less than that of water, air cannot move as large a particle as water flowing at the same velocity. In most regions with moderate to strong winds, the largest particles that can be lifted by the air are grains of sand. Saltation of sand grains accounts for the majority of sediment transport by wind; only the finest dust particles remain aloft long enough to be moved by suspension.

Flowing air erodes the land surface in two ways. The first, **abrasion,** results from the impact of wind-driven grains of sand (Figure 6.4). Airborne particles act like tools, chipping small fragments off rocks that stick up from the surface. When rocks are abraded in this way they acquire distinctive, curved shapes and a surface polish. A bedrock surface or stone that has been abraded and shaped by wind-blown sediment is called a **ventifact** ("wind artifact"). The second wind erosional process, **deflation** (from the Latin word meaning "to blow away"), occurs when the wind picks up and removes loose particles of sand and dust (Figure 6.5). Deflation on a large scale takes place only where there is little or no vegetation and loose particles are fine enough to be picked up by the wind. It is especially severe in deserts, but can occur elsewhere during times of drought when no moisture is present to hold soil particles together.

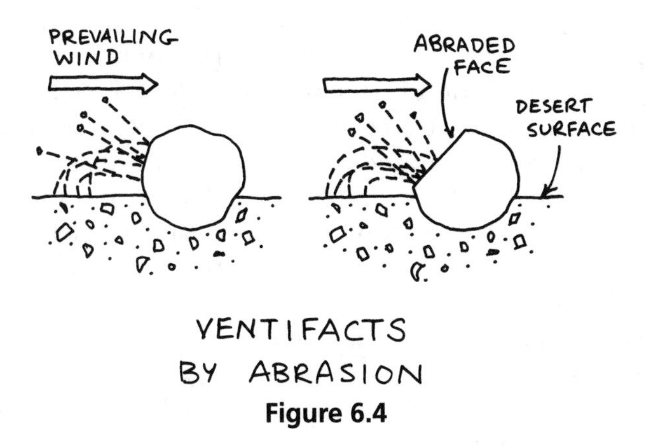

Figure 6.4

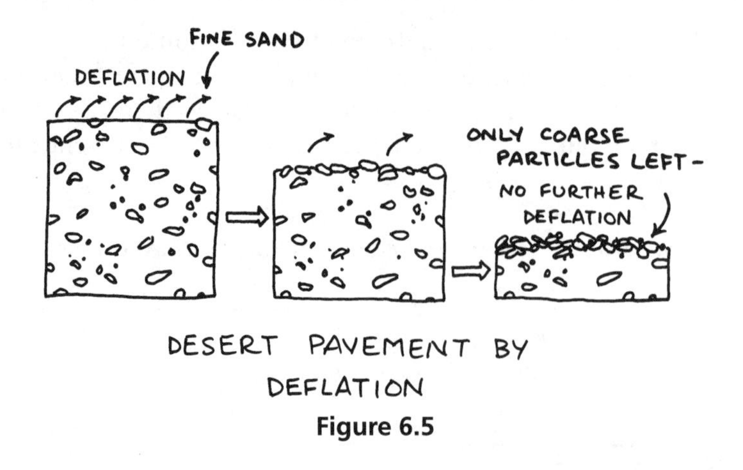

Figure 6.5

What are the two mechanisms by which wind erodes the land surface?

Answer: Deflation and abrasion.

9 EROSION BY ICE

A large body of ice that survives from year to year and shows evidence of movement is called a **glacier.** Glacial ice is a solid, but it does flow—usually very slowly—under the influence of gravity. As glaciers move, they sculpt the landscape and transport and deposit large volumes of material, creating distinctive glacial landforms.

Glaciers vary considerably in shape and size. The smallest, **cirque glaciers,** occupy protected, bowl-shaped depressions that are produced by glacial erosion on a mountainside. A cirque glacier that expands outward and downward becomes a **valley glacier.** Many high mountain ranges contain valley glaciers tens of kilometers long. A **piedmont glacier** is a broad lobe of ice, fed by one or more valley glaciers, that terminates on open slopes beyond a mountain front. Huge continent-sized **ice sheets,** found today only in Greenland and Antarctica, may reach thicknesses of more than 4 km (2.5 mi). Floating **ice shelves** hundreds of meters thick occur along the coasts of Antarctica, and smaller ones are found among the Canadian Arctic Islands. Approximately two-thirds of the Earth's ice is sea ice, floating on the ocean surface.

The size (or mass) of a glacier is a balance between the amount of snowfall in the winter and the amount of snow lost through melting and evaporation during the summer. Annual snowfall is generally very low in polar regions because the air is too cold to hold much moisture. The snow that does fall may not melt, how-

VALLEY GLACIER. The Yanert Glacier, Alaska, is an example of a valley glacier. Note the small valley glaciers entering from the right and becoming part of the larger glacier. (Courtesy U.S. Geological Survey)

ever, because summer temperatures are very low. Where the amount of snow that falls in winter is greater than the amount that melts during the following summer, the covering of snow gradually grows thicker. As it accumulates, the increasing weight causes the snow at the bottom to recrystallize into a more compact form, turning it into ice (Figure 6.6). Additions to the glacier's mass are called **accumulation,** and losses are called **ablation.** Over a period of years, a glacier may gain more mass than it loses; in such cases, the glacier's front, or **terminus,** is likely to advance. Conversely, a succession of years in which a glacier's mass decreases will cause the terminus to retreat.

Figure 6.6

Measurements of surface velocity in valley glaciers show that the uppermost ice in the central part of the glacier moves faster than the ice at the sides. Typical flow velocities range from a few centimeters to a few meters a day, about the same rate at which groundwater percolates through crustal rocks. Glacial ice moves in two basic ways: by internal flow and by basal sliding across the underlying rock or sediment. As the weight of overlying snow and ice increases, ice crystals deep within the glacier are deformed by slow displacement along internal crystal planes, like cards in a deck of playing cards sliding past one another (Figure 6.7). As the compacted, frozen mass moves, the crystal axes of the individual ice crystals are forced into the same orientation and end up with their internal crystal planes oriented in the same direction.

Basal sliding, the other type of movement in glaciers, occurs when ice at the bottom of the glacier slides across its bed. Meltwater at the base may act as a lubricant; in such glaciers, basal sliding can account for up to 90 percent of total observed movement. However, some glaciers are so cold that they are frozen to their bed, so they must move primarily by internal flow. The base of a glacier is studded with rock fragments of various sizes that are all carried along with the moving ice. When basal sliding occurs, small fragments of rock embedded in the ice scrape away at the underlying bedrock and produce long, nearly parallel scratches called **glacial striations** and **glacial grooves.** Because they are aligned parallel to the direction of ice flow, glacial striations and glacial grooves help geologists reconstruct the flow paths of former glaciers.

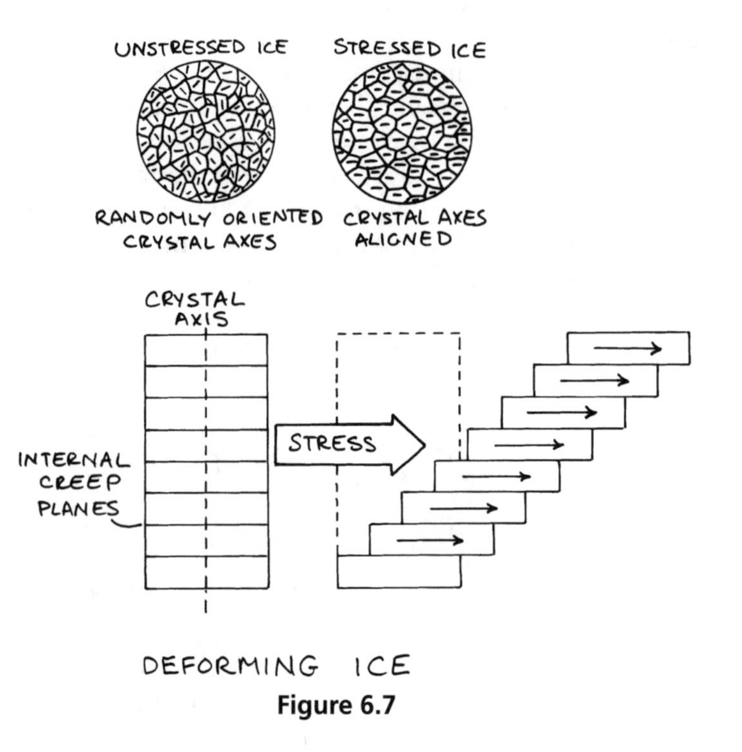

Figure 6.7

Occasionally glaciers exhibit an unusual phenomenon called a **glacial surge,** marked by rapid movement and dramatic changes in size and form. The rate of movement during a surge can be up to 100 times that of a nonsurging glacier. The causes of surging are poorly understood, but may be related to a buildup of meltwater at the base of the glacier that reduces the normal friction and permits unusually rapid basal sliding.

As glaciers move, they change the landscape by eroding and scraping away material and by transporting and depositing material at their ends and along their margins. A glacier erodes the land by acting like a file, a plow, and a sled. As a file, it rasps away firm rock. As a plow, it scrapes up weathered rock and soil and plucks out blocks of bedrock. As a sled, it carries away the load of sediment acquired by plowing and filing, along with rock debris that falls onto it from adjacent slopes. Let's look at the landforms that result from these processes.

In mountainous regions, glaciers produce a variety of distinctive landforms, such as bowl-shaped **cirques,** which are formed through a combination of plucking, frost wedging, and abrasion. As cirques on opposite sides of a mountain are eroded, their walls meet to form a sharp-crested ridge called an **arête.** When glacial ice moves downward from a cirque, it will carve a distinctive U-shaped valley channel. The retreat of a glacier can leave behind a terrain full of scoured pits and pockmarks, which fill with water to form ponds and lakes. An example of a small lake formed by glacial processes is Walden Pond, made famous by the writer

GLACIAL EROSION. A valley once occupied by a glacier has smooth sides and a characteristic U shape, created by the scraping action of the glacial ice as it flowed slowly down the valley. The view is from the head of Deadman Canyon, Tulare County, California. (Courtesy U.S. Geological Survey)

Henry Thoreau. The Great Lakes, Lake Winnipeg, and Great Bear Lake are examples of very large, glacially formed lakes.

Like streams, glaciers carry a load of sediment particles of various sizes. Unlike a stream, however, a glacier can carry part of its load at its sides and even on its surface. A glacier also can carry very large rocks and small fragments side by side without separating them by size or weight. Thus, sediments deposited by a glacier are not sorted or stratified the way stream deposits usually are. This can be seen by

GLACIAL DEPOSITS. Glacial till is an unsorted jumble of coarse and fine debris deposited by a glacier. This till was photographed at Pine Rock, New Haven, Connecticut. (Courtesy Brian Skinner)

examining glacial **till,** a heterogeneous mixture of finely crushed rock, sand, pebbles, cobbles, and boulders deposited by a glacier. In most cases, the boulders and rock fragments are different from the underlying bedrock, indicating that the components of the till were transported to their present site from somewhere else. A glacially deposited rock that is different from the underlying bedrock is called an **erratic,** from the Latin for "wanderer."

Underneath some large glaciers are flowing streams that carry meltwater and sediment. When such streams emerge from the terminus of the glacier, they may deposit their sediment load, forming a broad, sweeping plain called an **outwash plain.** If the glacier subsequently retreats, the former locations of the streams may be marked by sinuous, curving deposits of stream sediment called **eskers.** Another common glacial landform is the **drumlin,** a streamlined, elongate hill consisting of glacially deposited sediment that lies parallel to the direction of ice flow.

The boulders, rock fragments, and other sediment carried by the glacier may be deposited along its margins or terminus. A ridge or pile of debris that is being transported or has been deposited along the sides or terminus of a glacier is called a **moraine.** Geologists have used the locations of glacial moraines in the United States and Canada to determine how far the glacial ice cover extended over North America during the Ice Age of the Pleistocene epoch.

How does a glacier erode and sculpt the landscape?

____

____

____

Answer: As a file, it rasps away firm rock. As a plow, it scrapes up weathered rock and soil and plucks out blocks of bedrock. As a sled, it carries away the load of sediment acquired by plowing and filing.

10 MASS WASTING AND LANDSLIDES

All slopes are subject to mass wasting, the downslope movement of regolith under the influence of gravity. Exactly how and how fast the movement happens is controlled by the composition and texture of the regolith and bedrock, the amount of air and water present, and the steepness of the slope. For convenience, we can divide mass wasting into two categories (Figure 6.8).

1. **Slope failures** occur when relatively coherent masses of rock move downslope as the result of sudden failure on a steep slope or cliff.
2. **Flows** occur when loose mixtures of sediment, water, and air move downslope in a chaotic, fluid manner.

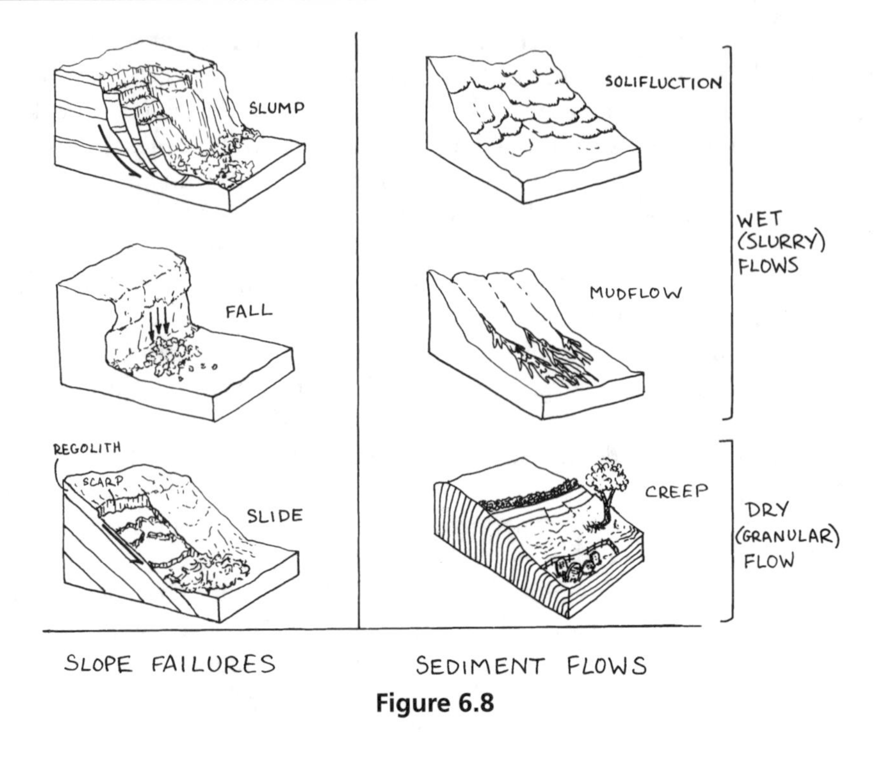

Figure 6.8

Slope failures include falls, slides, and slumps (see Figure 6.8). A **fall** is a sudden, vertical drop of rock fragments or debris. **Slides,** like falls, involve the rapid displacement of a mass of rock or sediment down a slope, usually along a flat (planar) surface. A **slump** is a type of slope failure that involves rotational movement of rock and regolith—that is, downward and outward movement along a curved surface. Slope failures can be triggered by natural causes, such as earthquakes, or by the oversteepening of a slope as a result of road- or house-building activities. The downslope movement of rock or regolith in a slope failure may be facilitated by the presence of excess water, but the main driving force is gravity.

Flows involve dense mixtures of regolith with water and/or air (see Figure 6.8). These mixtures are characterized by chaotic, fluidlike downslope movement. Water-saturated mixtures give rise to wet flows, or **slurry flows,** such as mudflows. Mixtures that are not water-saturated are called **granular flows.** Flows can be either slow or fast. The most common, widespread kind of mass wasting is **creep,** the imperceptibly slow, downslope, granular flow of regolith. Most of us have seen evidence of creep in curved tree trunks or old fences, telephone poles, or gravestones leaning at an angle on hillslopes.

Mass wasting is especially prevalent in **periglacial** landscapes, areas that are in close proximity to glaciers (*peri* means "near"). In such regions, intense frost action and a large annual range in temperature create a distinctive set of landforms.

A feature typical of periglacial terrains is **permafrost,** ground that is below the freezing point of water year-round. During the short summer, the frozen ground thaws in a thin layer near the surface. This leaves the soil water-saturated and vulnerable to a type of mass wasting called **solifluction,** the downslope movement of water-saturated regolith.

Slides and other forms of mass wasting cause extensive damage and loss of life each year. With careful analysis and planning, and appropriate stabilization techniques, the impacts of mass wasting can be reduced. Eliminating or restricting activities in areas where slides are likely to occur may be the best way to mitigate such hazards. For example, land that is susceptible to failure might be suitable for some types of development (recreation or parkland) but not others (intensive agriculture or housing). Some engineering techniques that can mitigate or prevent slope failure include retaining walls, drainage pipes, grading of slopes, and diversion walls.

A glacially deposited rock or boulder that is different from the underlying bedrock is called ___________.

Answer: An erratic.

In geology, as in many other sciences, the first step toward understanding is to learn the language of the science. It seems like there are endless new vocabulary words, but if you study the key words in this and the other chapters, you will soon be able to "talk the talk." The basic concepts will begin to fall into place as well. Now try the Self-Test for this chapter.

SELF-TEST

These questions are designed to help you assess how well you have learned the concepts presented in chapter 6. The answers are given at the end.

1. A slope failure that involves rotational movement of a coherent block of rock along a curved slip surface is called a ___________.
   a. rock slide
   b. slump
   c. granular flow
   d. fall
2. Chemical weathering is caused by water that is slightly ___________.
   a. warm
   b. alkaline
   c. frozen
   d. acidic

3. Stream discharge refers to the ___________.
   a. volume of water flowing past a point on the banks of a channel during a given period of time
   b. amount of sediment carried by a stream
   c. velocity of flow in a stream channel
   d. place where a stream enters a standing body of water and deposits its sediment load

4. A bedrock surface or stone that has been abraded and shaped by windblown sediment is called a(n) ___________.

5. Ground that is below the freezing point of water year-round is called ___________.

6. Glacial ice moves in two basic ways: ___________ and ___________.

7. Chemical weathering tends to dominate in tropical regions, whereas mechanical weathering is more common and more effective in colder regions. (T or F)

8. Air cannot move as large a particle as water flowing at the same velocity, because the density of air is much less than the density of water. (T or F)

9. Match each description with the name of a type of landslide.

a. The free falling of detached bodies of bedrock from a cliff or steep slope.	i. slump
b. Downward and outward rotational movement of rock or regolith along a curved surface.	ii. slide
c. The sudden and rapid downslope movement of detached masses of rock across an inclined surface.	iii. flow
d. A flowing mass of loose regolith that has enough water content to make it highly fluid.	iv. fall

10. What is the difference between mechanical weathering and chemical weathering? Give at least two examples of each.

____

____

____

____

11. Which common mineral is highly resistant to chemical weathering?

____

12. How do particles move by saltation?

____

____

____

ANSWERS

1. b
2. d
3. a
4. ventifact
5. permafrost
6. internal flow; basal sliding
7. T
8. T
9. a-iv; b-i; c-ii; d-iii
10. In mechanical weathering, rocks are broken up physically with no change in chemical composition. Examples include frost wedging; crystallization of salt in fractures; jointing; exfoliation; root wedging; and fire splitting. In chemical weathering, rocks and minerals decompose or are dissolved as a result of chemical and biochemical reactions that produce new minerals. Examples include hydrolysis and ion exchange; solution (dissolution); and oxidation.
11. quartz
12. In saltation, particles that are too heavy to be picked up by flowing wind or water are bounced along the bottom in a series of arc-shaped paths. Rolling or saltating grains dislodge other particles, which also bounce forward in arc-shaped paths.

KEY WORDS

ablation
abrasion
accumulation
acid rain
anthropogenic
arête
bed load
cave
chemical weathering
cirque
cirque glacier
creep
deflation
deposition
discharge
dissolution

dissolved load
drumlin
eolian
erosion
erratic
esker
exfoliation
fall
flow
frost wedging
glacial groove
glacial striation
glacial surge
glacier
granular flow
groundwater
hydrolysis
ice sheet
ice shelf
joint
karst
laminar flow
load
mass wasting
meander
mechanical weathering
moraine
outwash plain
oxidation
periglacial
permafrost
piedmont glacier
point bar
saltation
sinkhole
slide
slope failure
slump
slurry flow
solifluction
solution
stream
suspended load
suspension
terminus
till
turbulent flow
valley glacier
ventifact
weathering

7 Sediments and Sedimentary Rocks

To see a World in a grain of sand . . .

—William Blake

OBJECTIVES

In this chapter you will learn

- about the three basic kinds of sediment—clastic, chemical, and biogenic;
- how soils form and how they are classified;
- where and how different types of sediment are deposited;
- how sediments become sedimentary rocks.

1 SEDIMENT AND SEDIMENTATION

You learned in chapter 6 that regolith differs from place to place because its formation—the balance between mechanical and chemical weathering—is controlled in part by climate. Climate also influences how regolith is transported during erosion, whether by wind, water, or ice. Climate, along with other characteristics of the surrounding environment, also determines how and where sediments are deposited. As a result, sediments and sedimentary rocks preserve a geologic record, like an archive of past climates and environments. Reading these sedimentary archives of the Earth's past environments is not always easy, but it is a fascinating challenge.

BEDDING. Sedimentary layers of the Wasatch Formation have been exposed by erosion. This photograph was taken from the lookout point at the Bryce Canyon headquarters of Bryce Canyon National Park, Utah. (Courtesy U.S. Geological Survey)

Recall from chapter 3 that rock layering—stratification—results from the arrangement of sedimentary particles in distinct layers (strata). Strata differ from one another because of differences in the characteristics of the particles or in the way in which they are arranged. The layered arrangement of strata in a body of sediment or sedimentary rock is referred to as **bedding.** Each stratum, or **bed,** within a succession of strata can be distinguished from adjacent beds by differences in thickness or character. The top or bottom surface of a bed is a **bedding plane.**

When you look at an outcropping of sedimentary rock, one of the first things you see is the bedding. Indeed, it is the presence of bedding that provides the clue that the rock was once sediment. Look at individual beds, and it will become apparent that there are differences between them. The differences arise as a result of the formation, transportation, and deposition of the sediment, and from the processes by which the sediment was converted to rock. As a first step in deciphering the sedimentary archive, we need to sort out the evidence. We do this by separating sediments and sedimentary rocks into three groups based on the kinds of particles they contain.

The three groups of sediments are:

1. **Clastic sediment** (from the Greek word *klastos,* meaning "broken fragment"), formed from loose, fragmental rock and mineral debris produced during weathering. This debris is also called **detritus** (from the Latin for "worn down"), so clastic sediment is also known as **detrital sediment.**
2. **Chemical sediment,** formed by precipitation of minerals dissolved in lake water or seawater.
3. **Biogenic sediment,** composed mainly of the remains of plants and animals.

Can you think of a geologic term from a previous chapter that contains the Greek root *klastos,* "broken fragment"?

Answer: Pyroclast (chapter 5). There are lots of other geologic terms with the root *klastos,* such as bioclast, volcaniclastic, and clasticity; can you figure out what these terms mean?

Let's look briefly at these three basic types of sediment—what they are like, how they form, and how they are deposited.

2 CLASTIC SEDIMENT

Each individual particle in a clastic sediment is a **clast.** Clasts can be mineral grains or rock fragments. Clast shapes vary from angular to rounded, and they range in size from the largest boulders down to submicroscopic clay particles. Clast size is the primary basis for classifying both clastic sediments and clastic sedimentary rocks (Figure

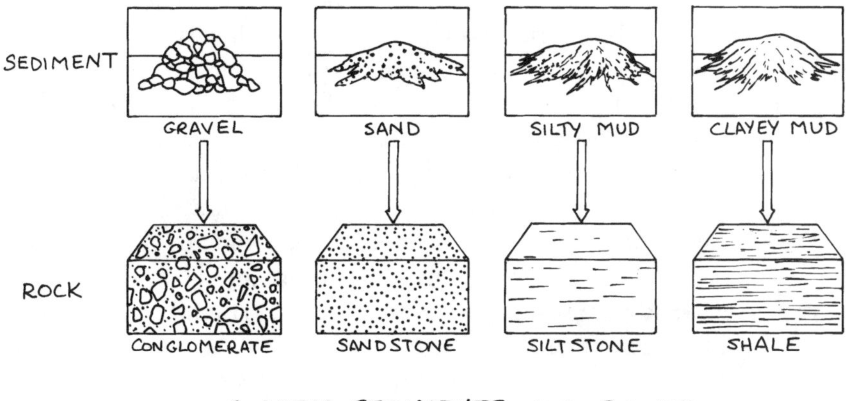

Figure 7.1

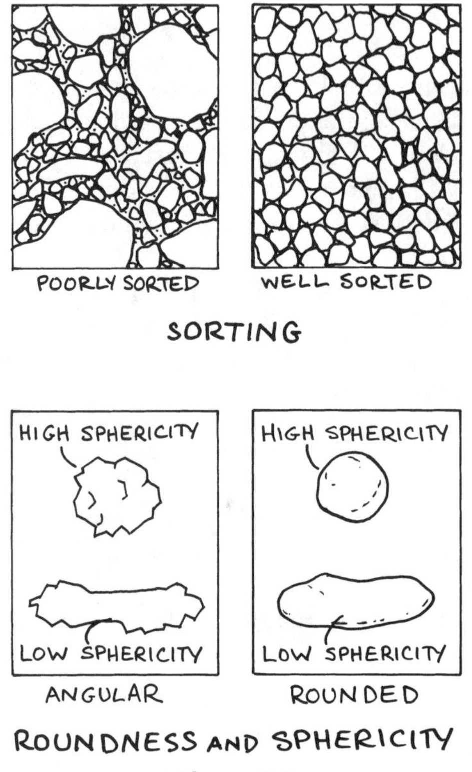

Figure 7.2

7.1). The four basic types of clastic sediment, based on the sizes of particles they contain, are **gravel** (pebbles, cobbles, and boulders); **sand** (particles from 0.06 to 2 mm [0.0025 to 0.08 in] in diameter); **silt** (particles from 0.004 to 0.06 mm [0.00016 to 0.0025 in] in diameter); and **clay** (particles less than 0.004 mm [0.00016 in] in diameter). The range of clast sizes within a given sediment reflects a characteristic called **sorting.** A poorly sorted sediment has a wide range of clast sizes; in a well-sorted sediment, the particles are all similarly sized (Figure 7.2).

Clast sorting and clast shape reflect the mechanisms of sediment transport and deposition. Mass-wasted sediment and ice-transported sediment tend to be poorly sorted, with angular or subangular (that is, not quite angular) clasts. Poorly sorted or nonsorted sediments that are ice-transported are called tills (see chapter 6). On the other hand, water- and wind-transported sediments tend to be well sorted, with rounded clasts of uniform size. As the clasts are transported, they are subject to continuous chemical and physical breakdown. After several times through the rock cycle, each time being subjected to weathering, erosion, and deposition, the result can be a sediment that consists almost entirely of rounded grains of quartz, the most resistant of the rock-forming minerals.

Depositional processes also influence the characteristics of clastic sediments. For example, **rhythmic layering** consists of alternating layers of coarse and fine clasts. Such an alternation suggests that some naturally occurring rhythm has influenced the transport and deposition of the sediment. A pair of rhythmic sedimentary layers deposited during the cycle of a single year is called a **varve** (Swedish for "cycle"). Varves are formed by seasonal variations in glacial lakes. In spring and summer, the inflow of glacial meltwater carries coarse sediment. In autumn and winter, the flow of meltwater ceases, the lake freezes over, and any fine sediment that remains in suspension slowly settles to form the fine-grained bed of the varve pair.

Graded beds are individual beds in which the coarsest clasts are concentrated at the bottom of the bed, grading up to the very finest at the top. Graded beds form from mixtures of coarse and fine clasts, such as might be carried by a stream in a flood. When the flowing slurry of water and sediment slows down, the coarse clasts settle quickly, followed successively by finer ones.

VARVES. These annual pairs of sedimentary layers were deposited in a glacial lake in Connecticut about 12,000 years ago. The lighter-colored layers consist of coarse sand grains; they were deposited during the summer months when melting was rapid and water flow was swift. The thinner, darker layers are wintertime deposits formed when the lake froze over and the suspension of fine clay particles settled out. (Courtesy Richard Foster Flint)

You can demonstrate graded bedding using the same experimental setup you used to demonstrate the principle of original horizontality in chapter 3. Fill a tub (preferably clear plastic) with water and a mixture of sediment of various sizes—some pebbles, some coarse and fine sand, and some mud. Stir vigorously, and wait. The pebbles will settle out first, followed by the sand. The fine silt from the mud will settle last. You should be able to see the variation in grain sizes through the clear sides of the container.

Turbulent flow in streams, wind, or ocean waves produces a type of bedding called **cross-bedding,** which refers to beds that are inclined with respect to a thicker stratum within which they occur. As they are moved along by wind or water, particles tend to collect in ridges, mounds, or heaps in the form of ripples, waves, or dunes. These migrate slowly forward in the direction of the current. Particles are carried up and over the top of the pile by the current, accumulating on the downcurrent slope. This produces beds that are inclined. The direction in

CROSS-BEDDING. These cross-bedded strata in the de Chelly Sandstone, Monument Valley, Arizona, were formed in sand dunes. Note that even though individual strata are not horizontal, the whole package of strata is close to horizontal. (Courtesy Tad Nichols, Museum of Northern Arizona)

which the cross-bedding is inclined tells us the direction of flow of the water or air currents at the time of deposition.

Volcanic (or **volcaniclastic**) **sediments** are a special kind of clastic sediment. What makes them special is that all of the clasts are volcanic in origin. As discussed in chapter 5, explosive volcanic eruptions blast out large quantities of fragmental material during an eruption. There is an old saying that explains what is unique about volcanic sediments—they are "igneous on the way up but sedimentary on the way down." Because the fragments are hot, each is called a pyroclast (*pyro,* from the Greek word for fire).

What is the difference between rhythmic layers and varves?

Answer: Rhythmic layers are alternating fine and coarse layers of sediment (or sedimentary rock). A varve is a pair of rhythmic layers deposited within a single year, reflecting the influence of seasonal variations on sediment and sediment deposition.

VOLCANICLASTIC ROCK. Tuff is a rock made up of fragments of igneous rock erupted from a volcano. (Courtesy Peabody Museum, Yale University)

3 CHEMICAL SEDIMENT

Recall from chapter 6 that all surface water and groundwater contains dissolved salts, which eventually find their way to the sea. No water on or in the Earth is completely pure and free from dissolved matter. When dissolved matter is precipitated from seawater or lake water, chemical sediment is the result. This can happen in two ways. The first is through biochemical reactions resulting from the activities of plants and animals in the water. For example, tiny plants living in seawater can decrease the acidity of the water, causing calcium carbonate to precipitate. Many limestones form in this manner. Chemical sediments can also be precipitated through inorganic reactions, such as the evaporation of lake water or seawater. For example, if an inland sea becomes increasingly shallow in a warm climate, the seawater may begin to evaporate. Any salts that were dissolved in the seawater will be left behind as a residue. When the water in a hot spring cools, minerals may precipitate from it. Similar chemical sediments are deposited in kettles and hot-water pipes in the home.

What is the source of the dissolved matter in surface water on the Earth?

Answer: The main source is chemical weathering of minerals in surface rocks.

4 BIOGENIC SEDIMENT

Biogenic sediment is composed of the remains of organisms. This may seem similar to chemical sediment formed by a biochemical reaction, but there is an important difference. Chemical sediment is material that precipitates from a solution without ever being part of an organism, although the precipitation may be induced by biochemical reactions. Biogenic sediment, on the other hand, consists largely or entirely of material that was once part of living organisms. It is made of the fossil remains of animals or plants. Solid fossils, such as shells and bones, often

BIOCLASTIC ROCK. A coquina is a bioclastic sedimentary rock composed of broken shell fragments cemented together with calcite. The specimen is from Saint Augustine, Florida. (Courtesy Peabody Museum, Yale University)

end up as broken fragments, or clasts. If a biogenic sediment is largely composed of biogenic clasts it is called a **bioclastic sediment.**

An important type of biogenic sediment is **ooze,** a fine, muddy sediment that accumulates on the deep ocean floor. Ooze may be calcareous (composed mainly of the carbonate-bearing remains of marine organisms such as corals) or siliceous (composed mainly of the silica-rich remains of tiny floating marine organisms). Another type of biogenic sediment consists of the accumulated organic remains of terrestrial (land-based) plant and animal matter. **Peat** is a sediment that consists largely of unconsolidated plant remains. Fossilized organic material can be trapped in most types of sediments; under certain circumstances, the organic material may be transformed into oil or natural gas.

What is the difference between a biogenic sediment and a bioclastic sediment? Give an example of each.

____

____

____

Answer: A biogenic sediment consists of the remains of plants and/or animals; for example, calcareous or siliceous ooze. A bioclastic sediment consists of clasts (fragments) of biogenic material; for example, limestone made of shell fragments.

5 SOIL

Soil is a special kind of sediment. Soil forms the uppermost part of the regolith, but it is not just ordinary sediment; it is sediment that has been altered by biochemical processes, such that it can support rooted plant life. Examine a bit of soil with a microscope or magnifying glass and you will see that it contains a lot of stuff besides bits of rock and mineral. You will find fragments of **humus** (partially decayed organic matter), some tiny insects or worms, and even some bacteria.

Soil is an interactive medium in which all parts play a role. Humus performs the function of retaining the chemical nutrients released by decaying organisms and by the chemical weathering of minerals. Humus is critical to soil fertility, the ability of the soil to provide nutrients such as phosphorus, nitrogen, and potassium to growing plants. All of the functions performed by organisms and other soil constituents are part of the continuous cycling of nutrients between the regolith and the biosphere. With its partly mineral, partly organic composition, soil forms a bridge between the lithosphere and the biosphere. Like liquid water, soil is one of the features that makes the Earth a unique planet.

Soil evolves gradually. When fully developed it consists of a succession of zones, or **soil horizons,** each of which has distinct physical, chemical, and biological characteristics. The sequence of soil horizons, from the surface down to the underlying bedrock, constitutes a **soil profile.** Soil profiles vary considerably, influenced by such factors as climate, topography, and rock type. However, certain kinds of horizons are common to many profiles (Figure 7.3).

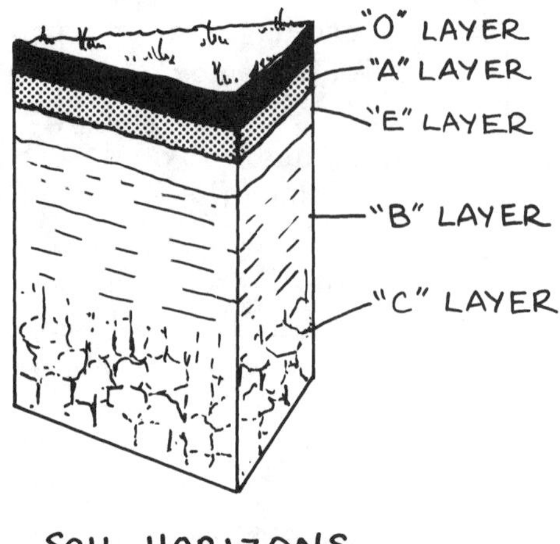

Figure 7.3

The uppermost horizon in many soil profiles is an accumulation of organic matter (O horizon). Below it lies the A horizon, dark in color because of the humus. The term "topsoil" is essentially a synonym for the O and A horizons. The E horizon, sometimes present below A, is typically grayish in color because it has little humus. E horizons are most common in acidic soils in evergreen forests. The B horizon underlies the A (or E) horizon. B horizons are brownish or reddish in color due to enrichment in clay minerals, iron and aluminum hydroxides transported downward from the horizons above. The C horizon, consisting of parent rock material in various stages of weathering, is deepest.

Because the kind of soil that develops in a given place depends on many variables, it is hardly surprising that classifications used by soil scientists are very complicated. The classification scheme used in the United States and many other countries is a hierarchical one, headed by 11 soil orders. Each order is distinguished by easily recognizable characteristics such as the presence or absence of well-developed horizons; accumulation of aluminum, iron, clay, or carbonate minerals; or high acidity and organic content. The 11 soil orders are divided into suborders and then, in increasing detail, into groups, subgroups, families, series, and types. Each of these has a name; no wonder soil classification is so complicated!

What is topsoil?

Answer: O (organic) and A (humus-rich) soil horizons.

6 LITHIFICATION AND DIAGENESIS

Lithification (from the verb "lithify," meaning "turn to stone") is the group of processes by which newly deposited, loose sediment is slowly converted to sedimentary rock. In order for this to happen, the clasts or particles must somehow be bound together. The chemical, physical, and biological changes that occur in sediments before, during, or just after lithification are referred to as **diagenesis.** Collectively, diagenetic changes turn sediment to rock.

The first and simplest diagenetic change is **compaction,** which occurs as the weight of an accumulating sediment forces the grains together. As the pore space (the space between grains) is reduced, water is forced out of the sediment. Compaction reduces the volume of the sediment considerably; it is most effective in sediments in which the grains can be closely stacked together. Thus, the sediments that are most highly compressible are those that consist of tiny, flat grains of silt or clay; sand, with rounded grains of hard quartz, is typically much less compressible.

As the pore water is squeezed out of a compacting sediment, substances dissolved in the water precipitate and cement the grains together. This is called **cementation.** Calcium carbonate is one of the most common cements. Silica, a particularly hard material, may also cement grains together. In the presence of oxygen (that is, in an oxidizing environment), ferrous iron in solution is oxidized and precipitated as ferric hydroxide; this also serves to bind sediment grains together.

As sediment accumulates, less stable minerals may recrystallize into more stable minerals. This process—**recrystallization**—is especially common in porous limestones that are formed from coral reefs. The mineral aragonite ($CaCO_3$) is present in the skeletal structures of living corals and other marine creatures. Over time, aragonite recrystallizes to calcite, a more stable form of calcium carbonate. Like cementation, recrystallization acts to hold grains together.

Which is more compressible—a silty sediment or a sandy sediment—and why?

____

____

Answer: A silty sediment would be more compressible, because the tiny, flat silt particles can be stacked closely together. A sandy sediment would contain lots of rounded grains of hard quartz, which would not compress easily.

7 CLASTIC SEDIMENTARY ROCKS

Clastic sedimentary rocks are classified on the basis of particle size, just as sediments are. The four basic classes are **conglomerate, sandstone, siltstone,** and **mudstone** (or **shale**). These are the rock equivalents of gravel, sand, silt, and clay (see Figure 7.1). A rock with large clasts is called a conglomerate if the clasts are rounded, but a **breccia** if they are angular. Angular clasts means that the sediment has only been transported a short distance and was not subjected to a long abrasion process. Most of the clasts in sandstones are quartz because quartz is such a resistant mineral. However, if the sediment has not been transported very far and the rock still contains a lot of feldspar (not as resistant as quartz), we use the term **arkose.** If lots of rock fragments are also present, the sandstone is called a **graywacke.** In rocks with the smallest clast sizes, a distinction is drawn between shales, which break into sheetlike fragments, and mudstones, which break into blocky fragments.

We give special names to clastic sedimentary rocks that form by certain special processes. For example, a rock that forms from cementation of a poorly sorted, glacially deposited till is called a **tillite.** A rock that forms through cementation or welding of hot pyroclastic fragments is called a **tuff** if the fragments are very fine, and **agglomerate** if the pyroclasts are coarse.

If a sedimentary rock is made up of particles derived from the weathering and erosion of igneous rock, how can we tell that it is sedimentary rather than igneous? In addition to such obvious clues as bedding, the rock texture provides evidence. Mineral grains in igneous rocks are irregular in shape, but clasts in sedimentary rocks commonly are rounded. The clasts may show signs of the abrasion they received during transport. Also, clastic sedimentary rock contains cement holding the particles together, whereas igneous rock consists of interlocking mineral grains without any cement. Fossils are another important feature for distinguishing between the two classes of rock. No organism can survive the high temperatures at which igneous rocks form, so the presence of ancient shells or evidence of plant life is an important clue to sedimentary origin.

What is the difference between an agglomerate and a volcaniclastic rock?

____

____

____

____

Answer: The distinction is subtle, but important. Both rock types contain volcanogenic fragments (fragments of volcanic origin). Agglomerates and tuffs are fundamentally igneous; they form as a result of the cementation or welding together of pyroclasts, often while they are still hot. Volcaniclastic rocks are truly sedimentary. They form as a result of the weathering of volcanic rocks and the subsequent erosion, deposition, and lithification of the volcanogenic sediment or igneous rock fragments.

8 CHEMICAL SEDIMENTARY ROCKS

Chemical sedimentary rocks result from the lithification of chemical sediments. Most chemical sedimentary rocks contain only one important mineral. That, together with the mode of precipitation, forms the basis for classification.

Chemical sedimentary rocks formed by evaporation of seawater or lake water are called **evaporites.** Examples of seawater evaporites are rock salt (the mineral halite, NaCl) and gypsum ($CaSO_4 \cdot 2H_2O$). Examples of lake water evaporites are sodium carbonate (Na_2CO_3) and borax ($Na_2B_4O_7 \cdot 10H_2O$). Many evaporite minerals are mined because they have industrial or human consumption uses. For example, most of the salt we eat and the gypsum used to make plasterboard is mined from evaporites.

You can grow your own crystals by evaporation. At a pharmacy, you can buy powdered alum. Dissolve as much alum as you can in a flat tray filled with hot water. Let the water evaporate over a period of days, and you will see small alum crystals begin to form.

Biochemically formed sedimentary rocks include certain kinds of limestones, made primarily of the mineral calcite ($CaCO_3$), and phosporites, which consist largely of the calcium phosphate mineral apatite [$Ca_5(PO_4)_3(OH,F)$]. Phosphorites, which are the principal source of phosphorus mined for fertilizers, form when the bones of fish and other dead marine animals dissolve in deep ocean water. The phosphorus-rich water is brought to the surface by the upwelling of deep ocean currents. Under surface conditions, apatite precipitates from the phosporus-rich water and rains to the bottom, where it accumulates.

The most economically important kind of chemical sedimentary rocks are **banded iron formations.** These unusual rocks are iron-rich, siliceous sediments that are entirely chemical in origin. They are the source of most of the iron mined today. All known banded iron formations are at least 1.8 billion years old. Banded iron formations are marine in origin. This means that the iron was once dissolved in seawater. Seawater today contains only slight traces of iron, because oxygen in the atmosphere oxidizes the iron to an insoluble (that is, not easily dissolvable) form. When the banded iron formations were formed there must have been a very large amount of iron dissolved in seawater. We can conclude, therefore, that when these rocks were formed there was very little oxygen in the atmosphere (because surface ocean water is in chemical equilibrium with the atmosphere). Why did the iron precipitate? One possibility is that microscopic organisms floating in the sea released oxygen as a result of photosynthesis, causing the dissolved iron to precipitate.

Give two examples of seawater evaporites and two examples of lake water evaporites.

Answer: Seawater evaporites include rock salt or halite (NaCl), and gypsum ($CaSO_4 \cdot 2H_2O$). Lake water evaporites include sodium carbonate (Na_2CO_3), and borax ($Na_2B_4O_7 \cdot 10H_2O$).

BANDED IRON FORMATION. This banded iron formation consists of iron-rich chemical sediment (dark) interbedded with siliceous chert layers (white). The outcrop is part of the Brockman Iron Formation, a 2-billion-year-old banded iron formation in the Hamersley Range of Western Australia. (Courtesy Brian Skinner)

9 BIOGENIC SEDIMENTARY ROCKS

Limestone is the most important of the biogenic rocks. Some limestone is biochemical in origin, but by far the majority of all limestones are bioclastic in origin. They consist of the fossilized shells, or shell fragments, of marine organisms. These organisms build their shells of carbonate, so limestones are formed chiefly of the carbonate mineral calcite. Under certain environmental conditions, calcite will be replaced atom by atom by the mineral dolomite [$CaMg(CO_3)_2$]; the resulting rock is called a dolostone.

Bioclastic limestones are often used as building stones in places like banks, malls, and public bathrooms. Next time you are out shopping, have a close look at the ornamental building stones around you, and see if you can spot the fragmental remains of shells in some of them (the light-gray or light-buff-colored rocks are most likely to be limestones).

A second very important class of biogenic rocks consists of the accumulated remains of trees, bushes, and grasses. As mentioned above, peat is a type of sediment that consists largely of organic material that has accumulated in a swampy environment. Eventually, peat may be lithified to form **coal.** Lithification—in this case it can be called "coalification"—involves compaction, release of water, and a slow increase in the carbon content of the coal. (For more on this process, see chapter 11.)

What is the difference between peat and coal?

____

____

____

____

Answer: Peat is a sediment that consists almost entirely of the accumulated remains of land plants. Coal is a rock; it is formed from peat that has been compacted and dewatered. (Most types of coal are sedimentary rocks, but some types of coal have been heated during lithification, to the point where they are considered to be metamorphic rocks.)

10 THE RECORD OF ENVIRONMENTAL HISTORY

Layers of sediment and sedimentary rock record the changing environmental patterns of the Earth's surface and the progress of life over more than 3 billion years. You have seen already that the size, shape, and arrangement of particles in sediments, as well as the geometry of sedimentary strata, provide us with evidence about the geologic environment in which sediments accumulate. These and other clues in ancient rocks enable us to demonstrate the existence of ancient oceans, coasts, lakes, streams, glaciers, swamps, and all the other places where sediments are deposited.

Fossils provide some of the most important clues about former environments. Some animals and plants are restricted to warm, moist climates, whereas others can live only in cold, dry climates. Using the climatic ranges of modern plants and animals as guides, we can infer the general character of the climate in which similar ancestral forms lived. For example, plant fossils can provide estimates of past rainfall and temperature for sites on land. Fossils of tiny floating organisms can tell us about former surface temperatures and salinity conditions in the oceans. Fossils are also the basis for telling the relative ages of strata; they have been very important in reconstructing the past 600 million years of Earth history (as discussed in chapter 3).

The color of fresh (that is, unweathered) sedimentary rock can also provide environmental clues. The color of a rock is determined by the colors of the minerals, rock fragments, and organic matter that compose it. Iron sulfides and organic detritus, buried with sediment, are responsible for most of the dark colors in sedimentary rocks; their presence implies deposition in a reducing (oxygen-poor) environment. Reddish and brownish colors result mainly from the presence of iron oxides, either as powdery coatings on mineral grains or as very fine particles. These minerals point to oxidizing conditions in the environment of deposition.

Irregularities formed by currents moving across a sediment, together with cracks, grooves, and other minor depressions, can be preserved on the bedding plane of a sandstone or a siltstone. For example, bodies of sand that are being moved by wind, streams, or coastal waves are often rippled, and such ripples may be preserved in the rock. Some mudstones and siltstones contain layers that are cut by polygonal markings. By comparing them with similar features in modern sediments, we infer that these are mud cracks, caused by shrinkage and cracking of wet mud as its surface dries. Mud cracks imply the former presence of tidal flats, exposed streambeds, desert lake floors, and similar environments. Footprints and trails of animals are often found with ripple marks and mud cracks. Even raindrop impressions made during brief, intense showers may be preserved in strata. All provide evidence of moist surface conditions at the time of formation.

How can the color of a sedimentary rock provide clues about the environment in which the sediment was deposited?

____

____

____

Answer: Dark colors are caused by iron sulfides and organic detritus; their presence implies deposition in a reducing (oxygen-poor) environment. Reddish and brownish colors are caused by iron oxides; their presence implies deposition in an oxidizing environment.

11 SEDIMENTARY FACIES

If you examine a vertical sequence of exposed sedimentary rocks, you will notice differences as you move upward from one bed to the next. The differences indicate that during deposition of that sequence of sediments, the environmental conditions in that location changed over time. If you follow a single bed sideways for some distance, you will again notice changes, indicating that at any given time during deposition, conditions differed from place to place.

This change in sediment character that takes place as we move from one depositional environment to another is referred to as a change of facies (pronounced "FAY-seez"). A **sedimentary facies** can be thought of as any sediment that can be distinguished from another sediment that accumulated at the same time, but in a different depositional environment. One facies may be distinguished from another by differences in grain size, grain shape, stratification, color, chemical composition, depositional structures, or fossils. Facies that are next to one another can merge, either gradually or abruptly. Coarse gravel and sand of a beach may pass very gradually offshore into finer sand, silt, and clay on the floor of the sea or a lake. Coarse, bouldery glacial sediment, on the other hand, may end abruptly against stream sediments at the margin of a glacier.

By determining the distinctive characteristics of different bodies of sediment or sedimentary rock, studying the relationships of different facies, and using these characteristics to identify original depositional settings, we can reconstruct a picture of the varied environments in a region during past geologic intervals.

Imagine a sequence of sedimentary rocks in which the lowest layer (#1) consists of dark, organic-rich siltstones; above this, layer #2 consists of sandstone; the next layer (#3) consists of calcareous mudstones and limestones with coral fragments; and the top layer (#4) consists of a deep-sea, calcareous ooze. What changes in depositional environment are suggested by this particular sequence of sedimentary facies?

____

____

____

____

Answer: The lowest layer (#1) suggests a muddy, organic-rich depositional environment, perhaps a swamp; layer #2 suggests a beach; layer #3 suggests shallow ocean water with coral reefs; and layer #4 suggests a deep-sea environment. The sequence overall seems to indicate a change in sea level in this location, starting with a swamp, progressing to nearshore and shallow water, and ending in a deepwater environment.

12 LAND DEPOSITIONAL ENVIRONMENTS

Streams are the main sediment transporters on land. Stream-deposited sediment differs from place to place depending on the type of stream, the energy available for transporting particles, the nature of the sediment load, and the depositional environment. Stream deposits form along channel margins, valley floors, mountain fronts, and the margins of a lake or an ocean (Figure 7.4). These are all places where changes in stream energy—and, therefore, changes in the stream's ability to carry its load—take place.

A large, smoothly flowing stream may deposit well-sorted layers of coarse and fine particles as it swings back and forth across its channel. If a stream is unable to move all the available load, it deposits the excess sediment as bars. The bars divide the flow and concentrate it into deeper channels on either side. The water repeatedly divides and reunites as it flows through interconnected channels separated by bars or islands. This is a **braided channel** (Figure 7.5). Large braided streams typically have many short-lived islands and constantly shifting channels. They tend to form in streams with variable discharge and easily eroded banks that can supply abundant sediment to the system.

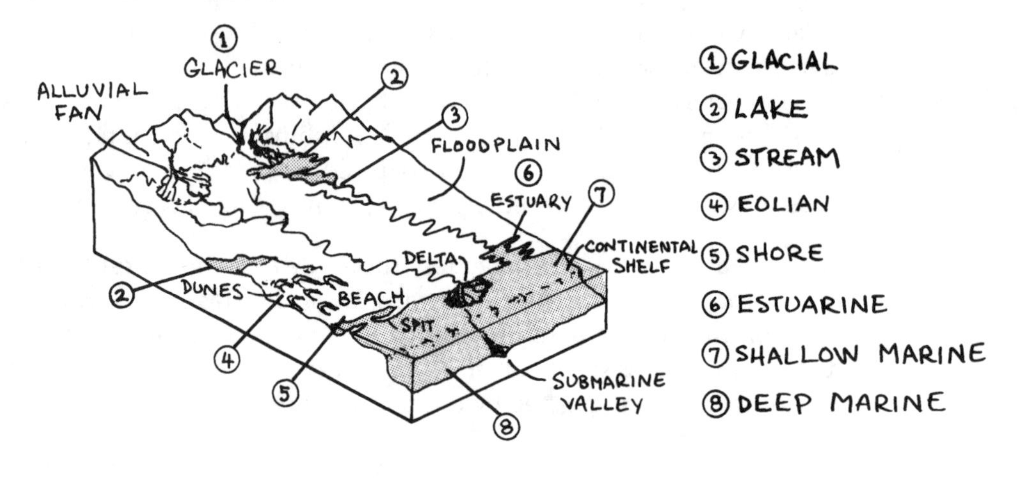

Figure 7.4

If a stream flowing through a steep upland valley suddenly emerges onto the floor of a much broader valley, it will experience a decrease in slope, velocity, and the ability to carry sediment. It will drop part of its load, forming a deposit called an **alluvial fan,** in which the sediments range from coarse, poorly sorted gravels to well-sorted sands (see Figure 7.4). A similar type of deposit that forms where a stream flows into standing water is a **delta,** so named because of its triangular shape, which resembles the Greek letter delta (Δ). When the stream enters the standing water, it quickly loses velocity and the heaviest particles drop out, forming a coarse, thick, steeply sloping layer. Most of the fine suspended load is carried farther seaward, eventually settling out to form a gently sloping delta front (see Figure 7.4). Most of the world's great rivers, including the Nile, the Ganges-Brahmaputra, the Huang He, the Amazon, and the Mississippi, have built massive deltas.

When a stream rises during a flood, water overflows the banks and inundates the adjacent valley floor, called the **floodplain** (see Figure 7.4). As sediment-laden water flows out of the channel, its depth, velocity, and turbulence decrease abruptly at the margins of the channel. This results in sudden, rapid deposition of the coarser part of the suspended load along the margins. Farther away, the finer particles settle out in the quiet water covering the valley, forming the broad, flat, fertile land that is typical of floodplains.

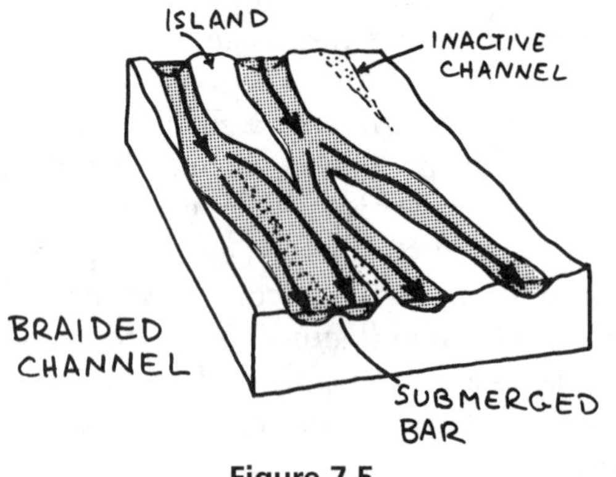

Figure 7.5

Sediment carried by the wind tends to be finer than that moved by

ALLUVIAL FANS. Large alluvial fans like these form where sediment-laden streams emerge from mountains onto a flat valley floor. (Courtesy U.S. Geological Survey)

other erosional agents. When windblown particles settle, they may form distinctive eolian deposits. Wind-laid dust that occurs in thick, uniform deposits is known as **loess** (the German word for "loose," pronounced "luhss"). Loess consists largely of silt, but is commonly accompanied by some fine sand and clay. In countries where it is widespread, loess is an important resource because of the productive (though highly erodible) soils that develop on it. The rich agricultural lands of the upper Mississippi Valley, the Columbia Plateau of Washington State, the Loess Plateau of China, and eastern Europe have loess soils that provide food for millions of people.

Sand grains are easily moved where strong winds blow and where vegetation is discontinuous, as along seacoasts and in deserts. In such places, the sand may pile up to form a hill or ridge composed of well-sorted sand grains, called a **dune** (Figure 7.6). Dunes develop where a minor surface irregularity or obstacle distorts the flow of air. The wind sweeps over and around the obstacle, but leaves a pocket of slower-moving air immediately downwind. In this pocket of low wind velocity, sand grains moving with the wind drop out and begin to form a mound. The mound, in turn, influences the flow of air over and around it and continues to

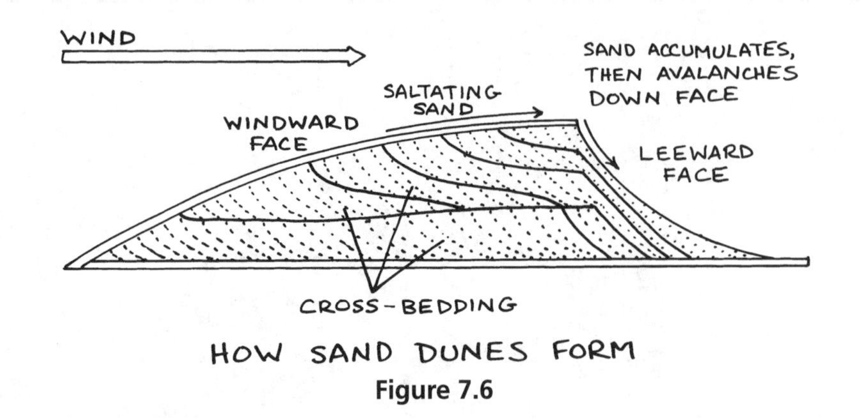

Figure 7.6

grow into a dune. A typical dune is asymmetrical, with a gentle windward slope (the side facing toward the wind) and a steep lee face (the side facing away from the wind). The fine bedding in sand dunes is typically inclined to the dominant, horizontal strata; this is called cross-bedding. Geologists can tell which way the wind was blowing from the cross-bedding and asymmetrical form of a dune.

On another sheet, draw and neatly label a diagram illustrating how dunes are formed.

Answer: See Figure 7.6.

13 MARINE DEPOSITIONAL ENVIRONMENTS

Streams transport detritus to the edges of the sea (see Figure 7.4). There it can accumulate near the mouths of streams, be moved laterally along the coast by currents to form a beach deposit, or be carried seaward to accumulate on the continental shelves.

Beaches consist of the coarsest particles contributed to the shoreline by the erosion of sea cliffs, or carried by rivers or currents moving along the shore. Quartz, the most durable of common minerals in continental rocks, is a typical component of beach sands. However, not all ocean beaches are sandy. Beach sediments tend to be well sorted, and typically display cross-bedding. Dragged back and forth by the surf and turned over and over, particles of beach sediment become rounded by abrasion.

Much of the load transported by a large river may be trapped in an **estuary,** a semienclosed body of coastal water where seawater mixes with freshwater (see Figure 7.4). Coarse sediment tends to settle close to the land, while fine sediment is carried in the seaward direction. Tiny individual particles of clay carried in suspension settle very slowly to the seafloor. When the sediment load is large, deltas may be built outward into the sea, as discussed previously. Large deltas are complex deposits consisting of coarse stream-channel sediments, fine sediments deposited between channels, and still finer sediments deposited on the seafloor.

Freshwater flowing from a river's mouth through an estuary will continue seaward across the submerged continental shelf as a distinct layer overlying the denser, salty marine water. Most of the coarse sediment is deposited within 5 to 6 km (3 to 4 mi) of the land; some fine-grained sediment, carried in suspension by the flowing water, reaches the outer shelf. This sediment then settles slowly to the seafloor. On the continental shelf of eastern North America, up to 14 km (almost 9 mi) of fine sediment has accumulated over the last 70 to 100 million years. Only about 10 percent of the sediment reaching the continental shelves remains in suspension long enough to arrive in the deep sea. The great bulk of the Earth's sedimentary strata is shelf strata whose sediment originated on the continents.

Thick bodies of sediment of continental origin lie at the foot of the continental slope in water depths as great as 5 km (3 mi). The origin of these sediments at great depths in the oceans was difficult to explain until marine geologists demonstrated that they could be deposited by **turbidity currents.** These are turbulent, gravity-driven flows consisting of dilute mixtures of sediment and water. The mixtures have a density greater than the surrounding water, so they rush swiftly down the offshore slopes at velocities up to 90 km/h (56 mi/h). As the current reaches the ocean floor it slows down and loses energy, depositing a graded layer of sediment called **turbidite.** At any given site along a continental rise, turbidites are deposited very infrequently, perhaps only once every few thousand years. Over millions of years, however, turbidites can slowly accumulate to form vast deposits beyond the continental realm.

Carbonate sediments of biogenic origin accumulate on the continental shelves wherever the influx of land-derived sediment is minimal and the climate and sea-surface temperature are warm enough to promote the abundant growth of carbonate-secreting organisms such as corals. In deeper water, biogenic calcareous ooze occurs over wide areas of ocean floor at low to middle latitudes. Other parts of the deep ocean floor are mantled with siliceous ooze from silica-secreting organisms, most notably in the equatorial Pacific and Indian Oceans, and in a belt encircling the Antarctic region. These are areas where surface waters have high biological productivity, in part related to the rise of deep-ocean water rich in nutrients.

Ocean water occupying a basin with restricted circulation will evaporate if the climate is warm enough. This leads to the precipitation of soluble substances and the accumulation of marine evaporite deposits. Such deposits are widespread. In North America, for example, marine evaporite strata of various ages underlie as much as 30 percent of the entire land area. An example of a modern-day environment in which an inland, freshwater sea is evaporating and becoming more saline is the Great Salt Lake in Utah.

Referring to the discussion about graded bedding in section 2 ("Clastic Sediment"), can you deduce why turbidites display graded bedding?

___

___

___

Answer: When turbidity currents rush downslope off the continental shelf, they are well-mixed slurries with a wide range of clast sizes. When they reach the bottom they lose velocity (and energy) very quickly, and drop their load. The coarsest grains settle first; the finest particles remain in suspension, eventually settling out on top. The result is a graded bed.

Now take the Self-Test to check your understanding and memory of the concepts and terms presented in this chapter.

SELF-TEST

These questions are designed to help you assess how well you have learned the concepts presented in chapter 7. The answers are given at the end.

1. Rhythmic layers are ___________.
   a. alternating fine- and coarse-grained layers in a sedimentary rock
   b. layers that are inclined with respect to a thicker layer within which they occur
   c. sedimentary layers that preserve ripple marks from wave action
   d. sedimentary layers that are extremely well sorted, with a high degree of sphericity

2. Phosphorites ___________.
   a. are chemical sediments that form as a result of biochemical processes
   b. form in a marine environment
   c. are the main source of phosphorous fertilizers
   d. All of the above are true.

3. An estuary is a ___________.
   a. semienclosed body of coastal water in which seawater mixes with fresh water
   b. complex of sediments consisting of coarse stream-channel deposits, fine sediments deposited between channels, and still finer seafloor deposits
   c. depositional environment that is typical of deep-ocean water
   d. fan-shaped deposit of sediments, ranging from coarse-grained to fine-grained

4. A well-sorted sediment contains clasts of many sizes and shapes. (T or F)
5. If a biogenic sediment consists mainly of fragments of biogenic origin, such as shells or bones, it is called a bioclastic sediment. (T or F)
6. Chemical sediment can be precipitated through ____________ or through ____________.
7. A conglomerate that consists primarily of angular clasts is a(n) ____________.
8. When sediments undergo compaction during lithification, the volume is reduced and the ____________ is forced out.
9. Fill in the blanks in the following table:

Sediment	Rock
gravel	____________
____________	sandstone
silt	____________
____________	mudstone

10. If a sedimentary rock is made up of particles derived from the weathering and erosion of igneous rocks, how can we tell that it is sedimentary rather than igneous?

____

____

____

____

11. What do ripple marks in a sedimentary rock reveal about the environment in which the sediment was originally deposited? What do mud cracks reveal about the environment in which the sediment was originally deposited?

____

____

____

12. What is deep-sea ooze?

____

____

13. What is "sorting" in a clastic sedimentary rock?

____

____

14. Describe the process of formation of a delta.

ANSWERS

1. a
2. d
3. a
4. F
5. T
6. biochemical reactions; inorganic reactions (such as evaporation)
7. breccia
8. pore water
9.

Sediment	Rock
gravel	conglomerate
sand	sandstone
silt	siltstone
clay	mudstone

10. From the (1) bedding (characteristic of sedimentary rocks but not igneous rocks); (2) rock texture (mineral grains in an igneous rock are irregular and interlocking, but clasts in a sedimentary rock are usually rounded); (3) cement (clastic sedimentary rock usually contains particles held together by cement, whereas igneous rock consists of interlocking mineral grains without any cement); or (4) presence of fossils (no organism can survive the high temperatures at which igneous rocks are formed).
11. Ripple marks reveal that the sediment was being gently moved by wind, streams, or waves, in a shallow-water environment. Mud cracks are caused by the shrinking and cracking of wet mud as its surface dries. They imply an intermittently wet environment, such as tidal flats, exposed streambeds, or desert lake floors.
12. Biogenic calcareous and siliceous materials, containing the remains or secretions of tiny sea creatures, which accumulate as muddy oozes on the seafloor.
13. Sorting refers to the uniformity of grain sizes in a clastic sedimentary rock. A well-sorted rock contains clasts that are all approximately the same size. A poorly sorted rock contains clasts of many different sizes.
14. When a stream flows into a large standing body of water, such as a lake, the stream water quickly loses velocity and the heaviest particles drop out, forming a coarse, steeply sloping layer. Most of the fine suspended load is carried farther out, eventually settling out to form a gently sloping, triangular deposit.

KEY WORDS

- agglomerate
- alluvial fan
- arkose
- banded iron formation
- beach
- bed
- bedding
- bedding plane
- bioclastic sediment
- biogenic sediment
- braided channel
- breccia
- cementation
- chemical sediment
- clast
- clastic sediment
- clay
- coal
- compaction
- conglomerate
- cross-bedding
- delta
- detrital sediment
- detritus
- diagenesis
- dune
- estuary
- evaporite
- floodplain
- graded bed
- gravel
- graywacke
- humus
- lithification
- loess
- mudstone
- ooze
- peat
- recrystallization
- rhythmic layering
- sand
- sandstone
- sedimentary facies
- shale
- silt
- siltstone
- soil horizons
- soil profile
- sorting
- tillite
- tuff
- turbidite
- turbidity current
- varve
- volcanic (or volcaniclastic) sediment

8 Metamorphism and Rock Deformation

In nature things move violently to their place, and calmly in their place.

—Sir Francis Bacon

It is useful to be assured that the heavings of the Earth are not the work of angry deities. These phenomena have causes of their own.

—Seneca

OBJECTIVES

In this chapter you will learn

- how and why rocks undergo deformation;
- how rock deformation creates structures such as faults and folds;
- how pressure, temperature, and fluids cause metamorphism in rocks;
- how metamorphic rocks and structures provide a geologic record of past tectonic events.

1 ROCK DEFORMATION AND PLATE TECTONICS

Plate tectonics tells us that lithospheric plates are constantly moving, colliding with one another and interacting along their margins. These interactions cause deformation of the Earth's crust. **Deformation** refers to all the different ways in which rocks respond to squeezing, stretching, or any other kind of tectonic force. When rocks deform, they may buckle and bend, crack and break, or flatten and change shape. It is easy to find evidence of this. If you look at a photograph of any great mountain range, you will see rock strata that were once horizontal and are now tilted and bent. Enormous tectonic forces are needed to deform such huge masses of rock. These forces come from the movement and interactions of lithospheric plates.

Most rock deformation happens slowly, within the crust or deep in the mantle, and we see the deformed rock only when it is uplifted and exposed at the surface by erosion. Therefore, we must infer how the deformation occurred long ago. Some of the evidence on which we base our inferences comes from laboratory studies; some comes from direct studies of deformed rocks.

Many of the landscapes we see around us are the results of rock deformation and the action of erosion on different rock strata. Nowhere is this more evident than in the great mountain ranges. Rock deformation is present in all the great mountain chains of the world, including the Alps, the Himalayas, the Urals, and the North American Cordillera (of which the Rockies are a part). Even in ancient

ROCK FOLDING AND DEFORMATION. These spectacular folded and deformed sedimentary rocks of Cambrian age are from the Sullivan River area in the southern Rocky Mountains in British Columbia, Canada. (Courtesy Geological Survey of Canada, Ottawa)

mountain chains like the Appalachians, which have been worn down by erosion, we can still see clear evidence of crustal deformation. Differences in the way adjacent strata erode have created distinctive topographic patterns that reveal the presence of deformed rocks underneath.

In what tectonic environment does the most intense rock deformation occur?

Answer: Along plate boundaries.

2 STRESS AND STRAIN

In discussing rock deformation, geologists often use the word **stress,** which refers to the force acting on a surface (per unit of area). The definition of **pressure** is exactly the same. However, the term stress often refers to "differential stress," that is, a situation in which the force acting on the surface of a body is greater from one direction than from another. Pressure is commonly used to mean "uniform stress," in which the force on a body is equal in all directions. For example, the pressure on a small body floating within a liquid is uniform—the same from all directions. Uniform stress is also called **confining pressure.** A rock in the lithosphere is confined by the rocks all around it and is uniformly stressed by those surrounding rocks. The related terms "lithostatic pressure" and "hydrostatic pressure" also describe uniform stress on a rock, but they convey additional information about how the pressure is transmitted to the rock: by overlying rocks (lithostatic, from *lithos,* the Greek root that means "rock"), or by water (hydrostatic, from *hydro,* the Greek root that means "water").

In response to stress, a rock will change its shape or its volume, sometimes both. This change is called **strain.** Uniform stress causes rocks to change their volume. For example, if a rock is subjected to uniform stress by being buried deep in the Earth, its volume will decrease. If the spaces (pores) between the grains become smaller, or if the minerals in the rock are transformed into more compact crystal structures, the volume change may be relatively large. Differential stress causes rocks to change their shape, and sometimes their volume as well.

There are several kinds of differential stress (Figure 8.1). **Tension** acts in a direction perpendicular to and away from a surface; this kind of stress pulls or stretches rocks. **Compression** acts in a direction perpendicular to and toward a surface; compressional stress squeezes rocks, shortening or squashing them and decreasing their volume. **Shear stress** acts parallel to a surface. It causes rocks to change shape by bending, flowing, or breaking. In response to shear stress, different parts of the rock may slide past each other like cards in a deck.

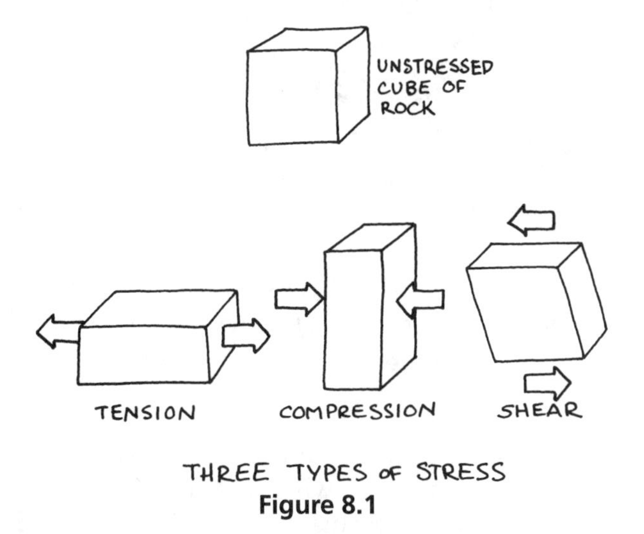

THREE TYPES OF STRESS
Figure 8.1

What is the difference between stress and strain?

Answer: Stress is a force acting on a rock body; strain is a change in the shape or the volume of a rock in response to stress.

3 TYPES OF DEFORMATION

When a rock is subjected to stress, it can respond in different ways. **Elastic deformation** is a nonpermanent change in the volume or shape of any solid, including rocks. When the stress is removed, the solid returns to its original shape and size. For example, if you stretch a metal spring and then let go, it will return to its original shape and size; this is elastic deformation. There is a degree of stress—called the elastic limit—beyond which the material is permanently deformed; that is, it does not return to its original size and shape when the stress is removed. If you stretch the metal spring too far, it won't return to its original shape; its elastic limit has been exceeded.

Try this yourself. Take a metal spring or ruler (one that you don't mind damaging). Bend the ruler or stretch the spring just a bit, and let it bounce back; that's elastic deformation (Figure 8.2). Now bend the ruler or stretch the spring, but keep going until you pass the elastic limit; the material has now undergone permanent deformation. What kind of stress were you applying? Did the material change its shape, its volume, or both?

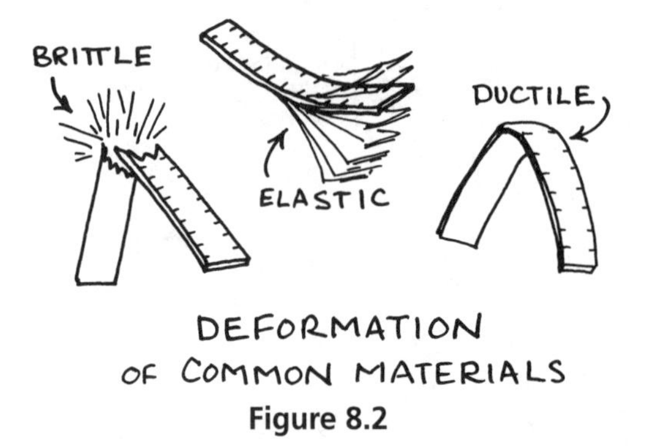

Figure 8.2

Rocks, too, will deform elastically up to a point. But if the stress continues and the elastic limit of the rock is exceeded, it will be permanently deformed. Under most circumstances, rocks can withstand only a small amount of elastic deformation before they deform permanently. **Ductile deformation,** also called "plastic" deformation, is one type of permanent deformation in a rock (or other solid) that has been stressed beyond its elastic limit. Rocks that deform in a ductile manner usually change their shape by flowing or bending.

Another type of permanent deformation is **brittle deformation,** in which the rock fractures instead of flowing or bending. A brittle material deforms by fracturing, whereas a ductile material deforms by changing its shape. Drop a piece of chalk on the floor and it will break. Drop a piece of cheddar cheese, however, and it will bend or squash instead of breaking. Under conditions of room temperature and atmospheric pressure, chalk is brittle and cheese is ductile. Similarly, some rocks behave in a brittle manner and others in a ductile manner.

Can you recall and relate the role of elastic deformation in earthquake generation (from chapter 4)?

____

____

____

Answer: In the elastic rebound theory of earthquake generation, rocks store elastic strain energy until they reach their elastic limit, at which point they fail suddenly, releasing the strain energy all at once and causing an earthquake.

4 FACTORS THAT INFLUENCE DEFORMATION

The higher the temperature, the more ductile and less brittle a solid becomes. At room temperature, for example, it is difficult to bend glass; if we try too hard, it will break because it is brittle. However, when it is heated over a flame, the glass becomes ductile and can easily be bent. Rocks are like glass in this respect. They are brittle at the Earth's surface, but they become ductile at great depths, where temperatures are high.

Confining pressure also influences deformation. High confining pressure reduces the brittleness of rocks because it hinders the formation of fractures. Experiments on cylinders of rock show that when a rock is supported by high confining pressure and subjected to compression from above, it will deform by bending or squashing rather than by fracturing. If a rock in low confining pressure is subjected to compression, it will tend to fracture. Rocks near the surface, where confining pressure is low, exhibit brittle behavior and develop many fractures. Deep in the Earth, where confining pressure is high, rocks tend to be ductile and deform by flowing or bending.

The rate at which stress is applied to a solid is another important factor in determining how the material will deform. If you take a hammer and whack a piece of ice suddenly, it will fracture. But if stress is applied to the ice little by little over a long period, it will sag, bend, and behave in a ductile manner. The same is true of rocks; if stress is applied quickly, the rock may behave in a brittle manner, but if small stresses are applied over a very long period, the same rock may behave in a ductile manner. The rate at which a rock is forced to change its shape or volume is called the **strain rate.** The lower the strain rate, the greater the tendency for ductile deformation to occur.

You can demonstrate the effects of different strain rates using Silly Putty, an unusual substance that can undergo both brittle and ductile deformation at room temperature. If you pull Silly Putty very slowly, it stretches into a long, rubbery string; this is ductile deformation. If you place a blob of Silly Putty on an inclined surface and check it an hour later, you will find that it has flowed partway down the surface; this, too, is ductile deformation. But if you pull Silly Putty apart very suddenly, it will break; this is brittle deformation caused by a rapid strain rate.

To summarize, low temperature, low confining pressure, and high strain rates tend to enhance the brittle behavior of rocks. Low-temperature and low-pressure conditions are characteristic of the crust, especially the upper crust. As a result, fracturing is common in upper-crust rocks. High temperature, high confining pressure, and low strain rates, which are characteristic of the deeper crust and the mantle, reduce the brittle properties of rocks. Fractures are uncommon deep in the crust and in the mantle because rocks at great depths behave in a ductile manner.

The composition of a material determines the temperature and pressure at which its transition from brittle to ductile behavior will occur. For example, both chalk and cheese behave in a brittle manner at 50 degrees below zero. At room temperature, cheese behaves in a ductile manner, while chalk is still brittle. They have different compositions and therefore different properties. The same is true of rocks and their mineral constituents. Some minerals, notably quartz, garnet, and olivine, are very brittle. In the crust, the transition from brittle to ductile behavior is controlled mainly by the deformational behavior of quartz-bearing rocks, such as granite. Other minerals, notably calcite, mica, clay, and gypsum, are more often ductile under natural conditions. Rocks such as limestone, marble, shale, and slate, which contain large quantities of these minerals, tend to deform in a ductile manner. The presence of water in a rock also enhances ductile properties, by reducing friction between mineral grains and by dissolving material at points of high stress.

What is the usual effect of high confining pressure on rock deformation?

Answer: In general, high confining pressure reduces the brittleness of rocks because it hinders the formation of fractures.

5 STRUCTURAL GEOLOGY

The study of stress and strain, the processes that cause them, and the types of rock structures that result from them are the subject of **structural geology.** Structural geologists study evidence of past rock deformation to find out what kinds of deformation have occurred in a particular area. This helps them understand the stresses that prevailed at different times in the past, which in turn allows them to decipher the geologic history of the area. Rock structures are also important from a practical perspective. For example, the locations of many types of ore deposits are controlled by the presence of structures such as faults. Rock deformation and the resulting structures can affect slope stability, influence the flow of groundwater, or trap oil and natural gas deep underground.

To understand and interpret rock structures, structural geologists make observations and measurements to determine what kinds of structures are present and specify the orientations and dimensions of those structures. They use these measurements and other sources of information to make inferences—educated guesses—about structures that may lie underground, hidden from view.

The first thing a structural geologist must do to study rock deformation is to determine the orientations of strata that have been tilted or otherwise deformed. The principle of original horizontality (see chapter 3) tells us that sedimentary strata are horizontal when they are first deposited. Where such rock strata are tilted, we can assume that deformation has occurred. The orientation of a rock layer is given by the **strike,** which is the compass direction (north, south, east, west) of the line of intersection between the rock layer and a horizontal plane (Figure 8.3).

We also need to measure the dip, the angle between the tilted rock layer and a horizontal plane. Dip is measured as an angle downward from the horizontal plane in degrees, using an instrument something like a protractor. If the rock layer is

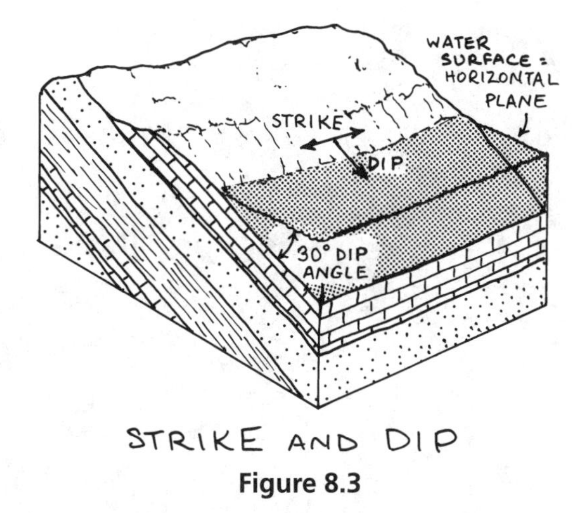

Figure 8.3

dipping shallowly, the angle of dip will be small; if it is dipping steeply, the angle will be greater. A vertical rock layer—one that has been tilted right up on its side—has a dip of 90° (it makes a right angle with the horizontal plane). Geologists use a symbol shaped like the following to indicate strike and dip on maps: ⟋⊤ . The orientation of the top bar of the symbol shows the strike, the bottom of the symbol shows the direction of dip, and a number tells the dip angle. Together, these three measurements—strike, direction of dip, and angle of dip—fully describe the orientation of a tilted rock layer. (The strike and dip symbols and other symbols commonly used on geologic maps are summarized in Appendix 4.)

What are some of the practical applications of structural geology?

____

____

____

Answer: From the text, locating and understanding the formation of ore deposits; assessing slope stability; predicting the flow of groundwater; and finding oil and natural gas buried deep underground. Can you think of any other possible applications?

6 FRACTURES AND FAULTS

Fractures, or cracks in rocks, are characteristic of brittle deformation. One important type of fracturing you are already familiar with (from chapter 6) is jointing. Because of the weight of overlying rock, rock masses buried deep beneath the Earth's surface are under enormous pressure. As erosion wears down the surface,

FAULT. This normal fault, in Jackass Butte, Idaho, has a dark layer of basalt with a lighter colored tuff underneath. A normal fault is a result of tensional forces. In this case the right-hand side of the fault has been displaced downward relative to the left-hand side. (Courtesy Brian Skinner)

the weight of the overlying rock, and hence the confining pressure, is reduced. The rock responds by expanding. As it does so, joints develop. Joints are a type of brittle deformation. They are different from faults because the blocks on either side have not moved significantly relative to one another along the fracture. A fault, as defined in chapter 4, is a rock fracture along which movement has occurred. There are different types of faults, caused by different kinds of stress. Faults are categorized on the basis of how steeply they dip and the relative movement of the rocks on either side of the fault.

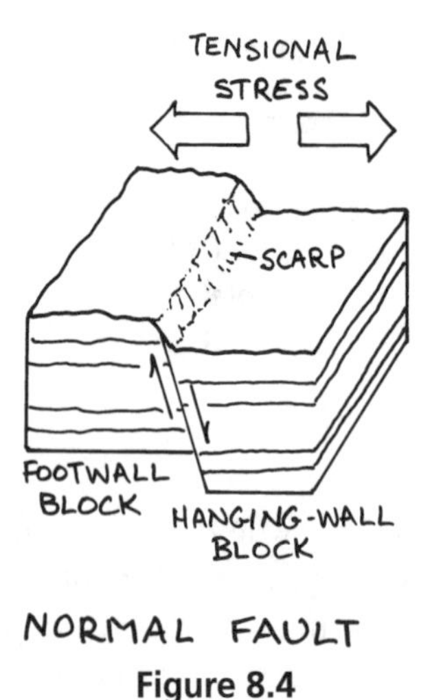

Figure 8.4

When tension stretches the crust, it causes **normal faults** (Figure 8.4). In a normal fault, the block of rock on top of the tilted fault surface (the "hanging-wall" block), moves down relative to the block on the bottom (the "footwall" block). Normal faulting allows the crust to lengthen and thin. Normal faults often occur in pairs. The block between the pair of faults will either drop down or pop up, depending on the direction of dip of the faults. In a **graben,** the two normal faults are dipping toward each other and the block between them drops down. In a **horst,** the two normal faults are dipping away from each other and the block between them is elevated. (If you look back at Figure 3.2 on page 51, you will see that it shows a set of horsts and grabens.)

These structures are common along divergent plate boundaries, where the crust is being stretched. The East African Rift Valley, for example, is a huge system of horsts and grabens marking the plate boundary along which the African continent is splitting apart.

When compression pushes blocks of crust together, **reverse faults** may result (Figure 8.5). In reverse faults the hanging-wall block is pushed up and over the footwall block, shortening and thickening the crust. When reverse faults are shallowly dipping (less than 30°), they are called **thrust faults.** Thrust faults are common in mountain chains along convergent plate boundaries. In large thrust faults, the hanging-wall block may move thousands of meters, coming to rest on top of much younger rock in the footwall block.

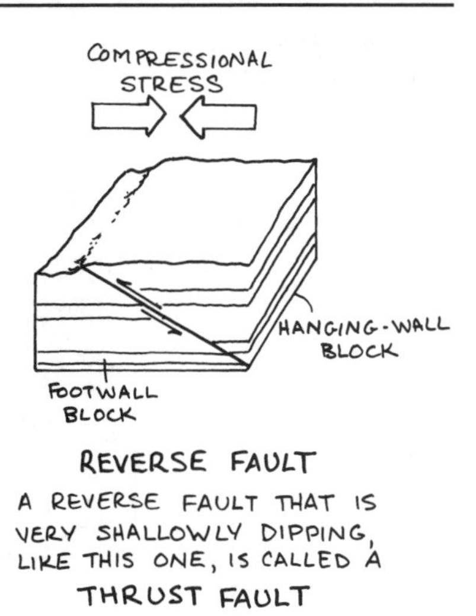

Figure 8.5

Strike-slip faults are caused by shear stress (Figure 8.6). The movement is mainly horizontal and parallel to the strike of the fault. One particularly famous strike-slip fault is the San Andreas Fault in California. Along this fault the Pacific Plate is moving toward the northwest relative to the North American Plate. Strike-slip faults are described according to the direction of relative motion of the blocks on either side of the fault. To an observer standing on either block, the movement of the other block is "left-lateral" if it has moved to the left and "right-lateral" if it has moved to the right. The San Andreas Fault is a right-lateral strike-slip fault. If you stand on the east side of the fault and look across toward the ocean, the block on the other side (the Pacific Ocean Plate) will be moving very slowly toward your right, that is, toward the northwest, carrying part of the California coast along with it. The relative motion in a right- or left-lateral fault is the same regardless of which block the observer is standing on. In other words, if you were to stand on the Pacific Ocean side of the San Andreas Fault and look across toward the continental United States, you would still have to turn your head to the right (toward the southeast) to track the movement of the other plate (the North American Plate) relative to where you are standing.

Figure 8.6

What are the three basic types of faults caused by tension, compression, and shear stress?

Answer: Normal, reverse, and strike-slip faults.

7 FOLDS

When rocks deform in a ductile manner, they bend and flow. The bending may be a broad, gentle warping over many hundreds of kilometers, or a tight flexing of microscopic size, or anything in between. Regardless of the size or shape of the structure, a bend or warp in a layered rock is called a **fold.**

The simplest type of fold is a **monocline,** a local steepening in otherwise uniformly dipping strata (Figure 8.7). An easy way to visualize a monocline is to lay a book flat on a table and drape a handkerchief over one side of the book; the draped handkerchief forms a monocline. An **anticline** is a fold in the form of an arch, with the rock strata convex upward. A **syncline** is a fold in the form of a trough, with the rock strata concave upward. Anticlines and synclines often occur together in sets (Figure 8.8). Sometimes a large area of crust undergoes upwarping or downwarping, which forms broad, gentle folds. Upwarping forms **domes,** while downwarping forms large, bowl-like **basins.**

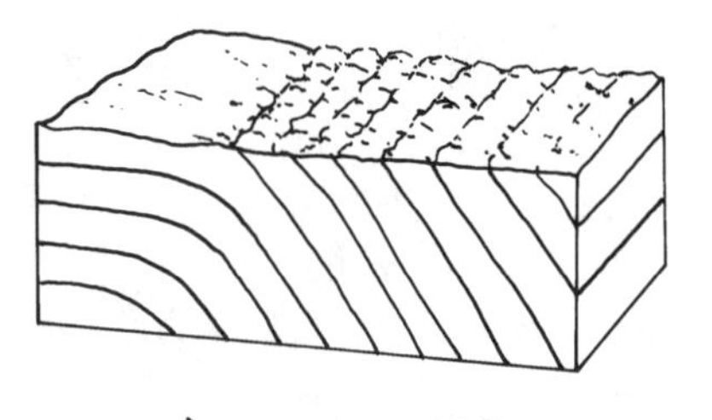

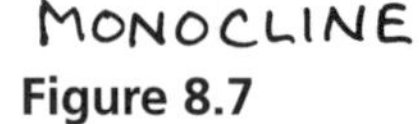

Figure 8.7

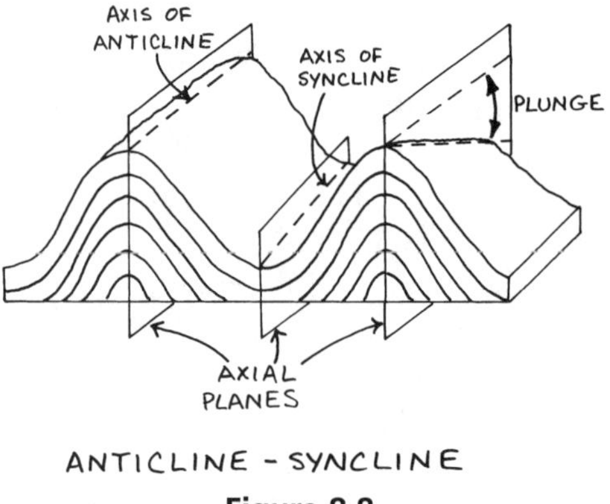

Figure 8.8

In a **symmetrical fold,** the two halves of the fold (the "limbs") are symmetrical relative to the fold axes. In an **asymmetrical fold,** one limb dips more steeply than the other. In an **overturned fold,** the bottom limb of the fold has been tilted beyond vertical so that it is upside down. Folds that are so strongly overturned that they are almost lying flat are called **recumbent folds.** Very tight folds with limbs that are nearly parallel are **isoclinal folds.**

Imagine a plane that divides a fold in half, as symmetrically as possible. This is the fold's **axial plane.** The line along which the axial plane intersects the rock strata in the fold is called the **fold axis.** The fold axis may be horizontal or it may be dipping, or "plunging." To fully describe the geometry and orientation of a fold and represent it on a map, we must define the following elements: (1) the orientation of the fold axis, shown as a line on the map; (2) the type of fold (for an anticline, we use two small arrows pointing away from the fold axis line, like this: ←|→; and for a syncline, we use two small arrows pointing toward the fold axis line, like this: →|←); (3) the direction in which the axis is plunging, indicated by an arrow on one end or the other of the fold axis line; and (4) the plunge angle (a number, in degrees).

Fold a piece of paper into (a) a monocline; (b) a syncline; (c) an anticline; and (d) a plunging anticline.

Answer: Refer to Figures 8.7 and 8.8.

8 GEOLOGIC MAPS

It is not possible to see all the structural details of deformed rocks in a given area because soil, water, vegetation, and buildings cover much of the evidence. Geologists gather information from **outcrops,** places where bedrock is exposed at the surface. We can also use information acquired from other sources, such as drill cores, to draw conclusions about what lies between the outcrops. The results of this information gathering are plotted on a **geologic map,** which shows the types, locations, and orientations of rock units and their structural features (Figure 8.9). Geologic maps help geologists interpret the geologic history of an area. They are used in a variety of ways by geologists from mining companies, oil companies, engineering firms, environmental agencies, and many others. Let's examine the techniques used to portray rock units and geologic structures in map format.

Objects portrayed on a map appear much smaller than they really are. It would be impossible, for example, to create a full-size map of North America. A map is a scale model, like a model train or car. We must specify the **scale**—the amount by which the size of objects shown on the map has been reduced. A scale of 1:1,000 ("1 to 1,000") means that one unit on the map is equal to 1,000 of the

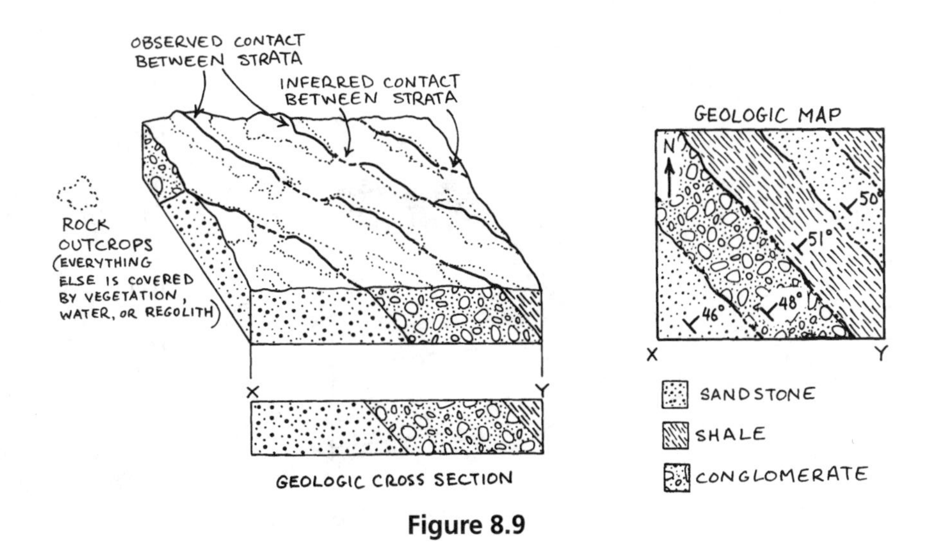

Figure 8.9

same unit on the Earth's surface. In other words, 1 km (1,000 m) would appear 1 m long on the map. To portray North America on a map of this scale would require a map more than 4 km (2.5 mi) wide! Therefore, geologic maps commonly use scales that reduce the sizes of objects even further. For example, 1:62,500 is a common scale for geologic maps in the United States; this scale is convenient because 1 inch on the map is equal to 1 mile on the ground. Where the metric system is used, as in Canada, Europe, and most U.S. government agencies, scales of 1:100,000 or 1:50,000 may be used.

A topographic map is commonly used as the base for a geologic map. **Topographic maps** show the shape of the ground surface and the location and elevation of features like valleys, hills, and cliffs. An important aspect of topography is **relief,** the difference between the lowest and highest elevations in the area. A mountainous region has high relief, whereas a flat plain has low relief. Topographic maps use **contour lines,** lines of equal elevation, to portray the shape and relief of the land surface, that is, its topography.

To understand contour lines, imagine a tank partly filled with water, with a clay model of a hill partly submerged by the water in the tank (as shown in Figure 8.10). The water makes a horizontal line—a line of equal elevation—around the hillsides. If you were to place a piece of glass over the tank and look down through the glass, you could trace the line onto the glass. Fill the tank with a bit more water (moving the horizontal line to a higher elevation) and trace the next line. Repeat the process, raising the water level by the same amount each time and tracing the resulting lines. You will end up with a set of topographic contour lines on the glass.

The contour lines represent the different elevations on the hillsides in this thought experiment, and the result is a flat, two-dimensional representation of

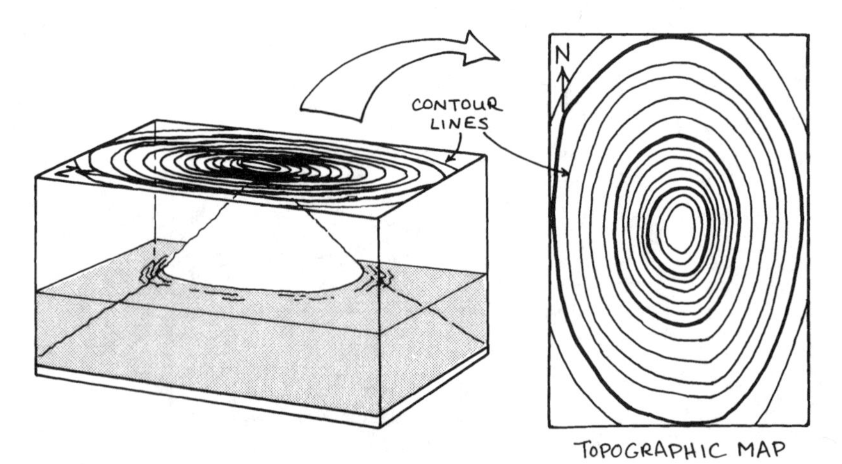

Figure 8.10

topography. The vertical difference—that is, the difference in elevation—between the successive water levels in the tank is called the contour interval. If you raised the water level by one centimeter each time, then the contour interval would be 1 cm. If you raised the water level by one inch each time, then the contour interval would be 1 in. On a real topographic map, the contour interval might be 10 ft, 10 m, or even 50 m. Where the land surface is steep, the contour lines are close together because the elevation changes a lot over a short distance. In contrast, where the land surface is flatter, the contour lines will be farther apart.

Geologists use a variety of colors, patterns, and symbols to portray different features on geologic maps. A **formation** is a unit of rock that can be mapped on the basis of rock type and recognizable boundaries, or geologic contacts, with other rock units. The colors and patterns used to portray rock formations on maps are standardized to a certain extent. Sedimentary rocks are often portrayed in green, blue, brown, or gray; recent (Quaternary) sediments in yellow; and igneous and metamorphic rocks in red, purple, or pink. A small dot pattern is typically used to represent sandstone, small parallel dashes for shales, irregular circles for conglomerate, a brick pattern for limestone, small V's for granite, wavy lines for metamorphic rocks, and so on.

A geologic map shows the locations of outcrops on the surface and the orientations of layering, structures, and other geologic features, as well as the geologist's educated guess as to what lies between the outcrops. Geologic maps can also be used to make inferences about what happens to the rock layers under the ground. Do they fold completely around and come back to the surface? Do they level out and become flat? Do they grade into a different type of rock? We need to know these things to figure out the geologic history of the area; to determine whether there are structures beneath the surface that could contain oil or mineral deposits; to assess the geologic stability of the area for construction purposes; and for many other reasons. In other words, we try to visualize the area in 3-D. We do this by constructing a **geologic cross section,** a diagram showing subsurface geologic features on an imaginary vertical plane (see Figure 8.9). Geologic cross sections are constructed on the basis of the information we have about the rocks at the surface, supplemented with information from drilling or other studies of the subsurface.

Why do close-set contour lines indicate steep topography, whereas contour lines that are farther apart indicate flat topography?

___

___

___

Answer: Each contour line represents an equivalent change in elevation. In steep topography, elevation changes a lot over a short distance, so contour lines are closely spaced. In flat topography, the elevation doesn't change very much, so contour lines are farther apart.

9 METAMORPHISM

Rock deformation is often associated with metamorphism. **Metamorphism** (from the Greek words *meta,* meaning "change," and *morphe,* meaning "form") encompasses all of the changes in mineral assemblage and rock texture that take place in the solid state as a result of changes in temperature and pressure. Metamorphic rocks preserve a record of the heatings, stretchings, and collisions that have happened to the crust throughout geologic history. Deciphering that record is an exceptional challenge for geologists.

Both mechanical deformation and chemical recrystallization processes are responsible for changes in the texture and mineral assemblages of metamorphic rocks. Mechanical deformation processes include grinding, crushing, and flattening. Chemical recrystallization processes include changes in mineral composition, growth of new minerals, and loss of pore fluids when a rock is heated and squeezed. Metamorphism always involves both mechanical deformation and chemical recrystallization, but their relative importance varies widely.

What are the two main groups of processes that cause metamorphic changes in rocks?

Answer: Mechanical deformation and chemical recrystallization.

10 THE LIMITS OF METAMORPHISM

Sediments and sedimentary rocks form in the upper 5 km (3 mi) of the crust; at a depth of 40 km (25 mi), near the base of the crust, melting and igneous processes start. In between—that is, through most of the crust—rocks remain solid but they slowly metamorphose, changing their character as a result of increasing pressure and temperature.

Pressure increases with depth in the Earth. At the surface the pressure is 1 atmosphere (atm). (Geologists often use bars or pascals [Pa] when referring to pressures within the Earth; 1 bar = 0.98692 atm = 10^5 Pa = 14.5 lb/m^2. The prefix *kilo,* abbreviated "k" as in km [kilometer], kg [kilogram], kbar [kilobar], or kPa [kilopascal], means "thousands." See Appendix 1 for more on units and conversions.) Pressure in the crust increases with depth at a rate of about 300 atm/km, or ≈ 300 bar/km. Therefore, at a depth of 5 km the pressure is 1,500 times greater than it is at the surface—that is, 1,500 atm (1.5 kbar). This represents the high-pressure limit of diagenesis and, with increasing pressure, the start of metamorphism. (Diagenesis—the set of processes by which sediment becomes rock—is discussed in chapter 7.)

We know from oil drilling and hard-rock mining that temperature in the continental crust also increases with depth, at a rate of 30°/km. At a depth of 5 km, the temperature is about 150°C (300°F). This is the temperature that separates diagenetic processes (below 150°C) from metamorphic processes (above 150°C). At the high-temperature end, the limit of metamorphism is about 800°C (almost 1,500°F); this is the temperature at which metamorphism ends and melting begins. These limits are approximate because pressure, temperature, and composition all interact to determine the onset of melting (as you may recall from chapter 5). Furthermore, metamorphism itself involves more than just temperature and pressure; for example, the presence of fluids such as water or carbon dioxide speeds up mineral reactions and also makes some reactions begin at lower temperatures.

Figure 8.11 may help you think about processes that happen deep inside the Earth. Pressure is represented on the vertical axis, increasing downward just as pressure increases with depth in the Earth. Temperature increases from left to right. The pressure and temperature conditions under which sediments form and are changed into sedimentary rocks by diagenesis occupy the upper left-hand corner (the low-temperature and low-pressure corner) of the diagram. From the upper limits of diagenesis to a temperature of about 550°C (just over 1,000°F) and a pressure of about 5 kbar (≈ 5,000 atm)—that is, a depth of about 18 km (11 mi)—metamorphism occurs but is not particularly intense. We refer to rocks that are formed under such conditions as **low-grade** metamorphic rocks. At temperatures and pressures above those of low-grade metamorphism, extending to the onset of melting at a depth of 30 to 40 km (19 to 25 mi), is the region in which **high-grade** metamorphic rocks form.

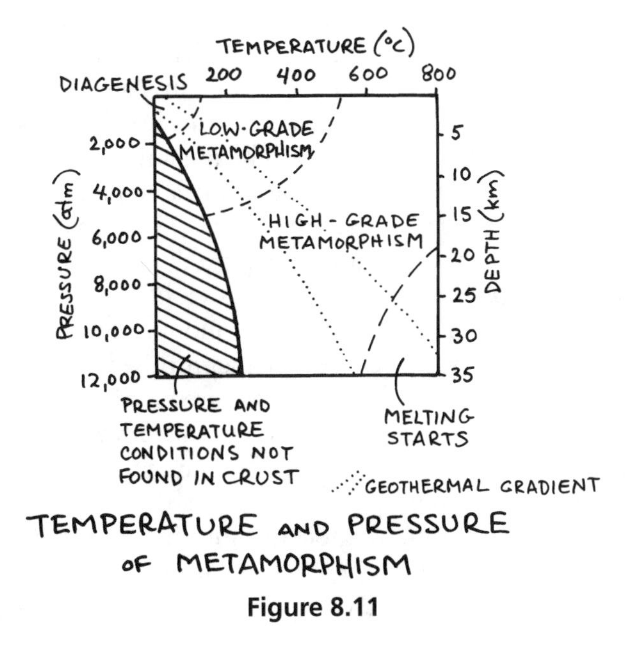

Figure 8.11

Can you recall (from chapter 5) what we call the normal change of temperature with increasing depth in the Earth? (Hint: it is also shown on Figure 8.11.)

Answer: The geothermal gradient.

11 FACTORS THAT INFLUENCE METAMORPHISM

Increasing temperature and pressure (stress) cause metamorphic changes in rocks. As temperature and stress increase, chemical reactions and textural changes occur; one new mineral assemblage follows another. For any given rock composition, each mineral assemblage is characteristic of a given range of metamorphic temperatures and pressures. The changes are also influenced by the presence or absence of fluids; how long the rock is subjected to high pressure or temperature; and whether the pressure was uniform or involved differential stress. Let's look briefly at the factors that influence metamorphism.

The pore spaces between the grains in a sedimentary rock and the tiny fractures in igneous rocks are commonly filled by a watery fluid. Pore water always has dissolved within it small amounts of gases and salts, plus traces of all the mineral constituents that are present in the enclosing rock. Pore fluids play an important role in metamorphism. They speed up chemical reactions in much the same way that water in a stew pot speeds up the cooking of a tough piece of meat. When pore fluids are absent or present only in tiny amounts, metamorphic reactions are very slow.

As pressure increases and metamorphism proceeds, the amount of pore space decreases and the pore fluid is slowly driven out of the rock. As temperature increases, hydrous minerals (minerals that contain water) begin to expel the water from their crystal structures. Water released in this way joins the pore fluid and is slowly driven out of the metamorphic rock. The metamorphic changes that occur while temperature and pressure are rising and while abundant pore fluid is present are collectively termed **prograde metamorphism.** Those that occur as temperature and pressure are declining, after much of the pore fluid has been expelled, are called **retrograde metamorphism.** Because of the lack of fluids, retrograde metamorphism happens less rapidly and its effects are less pronounced than those of prograde metamorphism.

One important temperature-controlled effect—the beginning of rock melting—is greatly influenced by the presence of pore fluid. The effect of pore fluid on rock is the same as the effect of salt on ice—it lowers the melting temperature. The upper temperature limit of metamorphism therefore depends on the amount of watery pore fluid present. When a tiny amount of pore fluid is present, only a small amount of melting occurs and the melt stays trapped in small pockets in the metamorphic rock. When the metamorphic rock cools, so do the little pockets of magma. The result is a composite rock—part metamorphic, part igneous—called a **migmatite.** When abundant pore fluid is present and large volumes of magma develop, the magma rises upward and intrudes the overlying metamorphic rock. As a result, we observe that batholiths (see chapter 5) tend to be closely associated with large volumes of metamorphic rock.

Textures in metamorphic rocks commonly record the effects of differential stress. The most striking effect of metamorphism under differential stress involves platy minerals, such as those in the mica family. Micas grow in flat sheets or plates. The plates line up parallel to one another, and perpendicular to the direction of maximum stress. The parallelism of platy mineral grains, along with other changes that happen during metamorphism, produce a distinctive planar texture called **foliation** (from the Latin word *folium,* meaning "leaf"). Low-grade metamorphic rocks tend to be so fine-grained that the newly formed mineral grains can be seen only with the microscope. The foliation is then called **slaty cleavage.** Grain sizes increase as a result of high-grade metamorphism so that individual mineral grains can be seen with the naked eye. Foliation in coarse-grained metamorphic rocks is called **schistosity** (derived from *schistos,* a Latin word meaning "cleaves easily").

How do migmatites form?

____

____

____

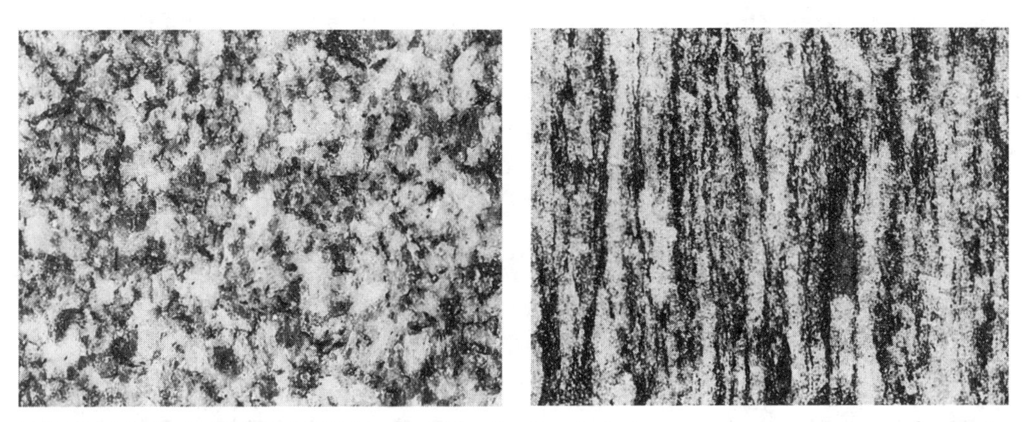

UNIFORM AND DIFFERENTIAL STRESS. On the left is a granite consisting of quartz (glassy looking), feldspar (white), and biotite (dark) that crystallized from a magma under conditions of uniform stress. Note that the biotite grains are randomly oriented. On the right is a metamorphic rock in which the mineral assemblage quartz + feldspar + biotite grew under differential stress. The biotite grains are parallel, giving the rock a pronounced layered texture. (Courtesy William Sacco)

ROCK CLEAVAGE. The rock cleavage developed in this folded rock from Vermont cuts across the original bedding. The cleavage results from the parallel growth of mica grains. (Courtesy Brian Skinner)

Answer: When a high-grade metamorphic rock approaches its melting point, small pockets of melt may begin to develop (partial melt; see chapter 5). If the rock lacks abundant pore fluid, the amount of melting will be small, and the melt may remain trapped in isolated pockets. When the rock cools, the small pockets of melt solidify. The resulting migmatite is part metamorphic and part igneous.

There are several kinds of metamorphism, distinguished from one another on the basis of the relative importance of mechanical deformation and chemical recrystallization and the tectonic environments and pressure-temperature conditions in which they occur. The most important kinds of metamorphism are contact metamorphism, burial metamorphism, and regional metamorphism.

12 CONTACT METAMORPHISM AND METASOMATISM

Contact metamorphism occurs near bodies of hot igneous rock that are intruded into cool sedimentary rocks of the crust. Such metamorphism involves mainly chemical recrystallization and happens in response to a pronounced increase in temperature; mechanical deformation is minor. The extent of contact metamorphism depends on the size of the intrusive body, the amount of pore fluid involved, and the kind of rock being metamorphosed. The zone ("aureole") of contact metamorphism around a small dike or sill may extend for only a few centimeters, especially if the dike or sill intrudes into a relatively unreactive rock. However, a large intrusion contains more heat energy than a small one and may also give off pore fluid as it cools. When intrusions are many kilometers across (stocks or batholiths) and when the intruded rocks are highly reactive (such as limestone), contact metamorphic aureoles may extend for hundreds of meters.

Where large quantities of fluids are present, as in a fracture through which fluids are flowing, the composition of rocks in the contact metamorphic aureole may be drastically altered. The term **metasomatism** (from *meta,* "change," and *soma,* from the Latin word for "body") refers to the process whereby rocks have their chemical compositions altered by the addition or removal of ions in solution. Metasomatism is commonly associated with contact metamorphism. The metasomatic fluids released by a cooling magma move outward, passing through the volume of rock undergoing contact metamorphism and altering its chemical composition. Rocks that are distant from the magma—beyond the reach of invading fluids—remain unchanged. Metasomatic fluids carry minerals in solution; if these minerals precipitate, they may form valuable mineral deposits.

What is the relationship between the size of a contact metamorphic aureole and the size of the intruding body?

____

____

____

Answer: If the intruding body is small, the contact metamorphic aureole will also likely be small. If the intruding body is large, it carries more heat energy; the aureole will also be large, especially if the host rock is highly reactive or a large amount of pore fluid is present.

13 BURIAL METAMORPHISM

Burial metamorphism is the first stage of metamorphism following diagenesis, the conversion of sediments to sedimentary rocks. Sedimentary rocks may reach depths of 10 km (about 6 mi) and temperatures of 300°C (572°F) or more when buried deep in a sedimentary basin. Metamorphic processes start at about 150°C (about 300°F) and are sped up by the abundant pore water that is always present in sedimentary rocks. But water-saturated rock is weak and acts more like a liquid than a solid. Therefore, the stress during burial metamorphism tends to be uniform. As a result, burial metamorphism involves mainly chemical recrystallization, with little mechanical deformation. Burial metamorphism is usually observed in deep sedimentary basins along the margins of tectonic plates.

What is the temperature that separates diagenetic processes from burial metamorphism?

Answer: 150°C.

14 REGIONAL METAMORPHISM AND METAMORPHIC FACIES

As temperatures and pressures increase, burial metamorphism grades into **regional metamorphism.** Regionally metamorphosed rocks are found in mountain ranges and in the eroded remnants of former mountain ranges. They form as a result of subduction or collisions between masses of continental crust. During a collision, rocks along the plate margins are subjected to very intense differential stress. The foliation that is characteristic of regionally metamorphosed rocks is a consequence of this stress.

To understand what happens during regional metamorphism, consider a segment of crust that is subjected to compression as a result of a continental collision. The rocks fold and buckle, causing the crust to become locally thickened. The bottom of the thickened mass of crust is pushed deeper into the mantle, where temperatures are higher. As a result, the rocks near the bottom of the thickened pile are subjected to both elevated stress and higher temperature. However, rocks are poor conductors of heat, so the heating-up process is slow. If the folding and thickening are very slow, as in the building of a great mountain chain such as the Himalayas, heating of the bottom of the pile may keep pace with the temperature of adjacent parts of the crust and mantle. However, if burial is very fast, as when rocks are dragged down in a subduction zone, the pile does not have time to heat up. Conditions of high pressure but rather low temperature will prevail in the down-going slab of rock. Depending on the rate of burial, therefore, the same starting rock can yield different metamorphic rocks because different pressure-temperature conditions were encountered.

Changes in mineral assemblage during metamorphism are determined principally by the temperature and stress to which the rocks are subjected. For a given rock composition, the assemblages of minerals that form under a given set of temperature and stress conditions are said to belong to the same **metamorphic facies** (as shown in Table 8.1). Each metamorphic facies represents a characteristic range of temperatures and pressures, and each facies is typical of metamorphism in a particular tectonic environment. (Compare this to the concept of sedimentary facies, discussed in chapter 7.)

Table 8.1 Pressure-Temperature Conditions and Metamorphic Facies

Pressure-Temperature Conditions	Facies Name	Type of Metamorphism	Typical Geologic Environment
Low temperature High pressure	Blueschist	Regional	Subduction zone
Low temperature Very high pressure	Eclogite		
Low temperature Low pressure	Greenschist		Continental collision zone
Moderate temperature Moderate to high pressure	Amphibolite		
High temperature High pressure	Granulite		
Very low temperature Very low pressure	Zeolite	Burial	Sedimentary basin
High temperature Low pressure	Hornfels	Contact	Contact metamorphic aureole

What are the two metamorphic facies that are characteristic of low-temperature, high-pressure metamorphism in a subduction zone?

____

Answer: Blueschist facies and eclogite facies.

15 METAMORPHIC ROCKS

The names of metamorphic rocks are based partly on texture and partly on mineral assemblage. The most widely used names are those applied to metamorphic rocks derived from shales, sandstones, limestones, and basalts. This is because shales, sandstones, and limestones are the most abundant sedimentary rock types, while basalt is by far the most abundant igneous rock. Let's look first at the naming of rocks with foliation and then rocks without foliation.

Slate forms as a result of low-grade metamorphism of shale, as shown in Table 8.2. The minerals in shale include quartz, clay, calcite, and feldspar. Under conditions of low-grade metamorphism, they recrystallize to form muscovite and/or chlorite. The tiny new mineral grains produce slaty cleavage, showing that the rock has gone from being a sedimentary rock to a metamorphic rock. Continued metamorphism produces larger grains of mica and a changing mineral assemblage. The rock develops pronounced foliation and is called **phyllite** (from the Greek, *phyllon,* "leaf"). In a slate it is not possible to see the new grains of mica with the unaided eye, but in a phyllite they are just large enough to be visible.

Table 8.2 Classification of Foliated Rocks

<table>
<tr><th rowspan="2">Original Rock</th><th rowspan="2">Metamorphic Rock</th><th rowspan="2">Texture</th><th colspan="2">Metamorphism</th></tr>
<tr><th>Kind</th><th>Grade</th></tr>
<tr><td rowspan="2">Shale</td><td>Slate</td><td rowspan="2">Fine-grained</td><td rowspan="7">Dominantly regional</td><td>Low</td></tr>
<tr><td>Phyllite</td><td rowspan="2">Medium</td></tr>
<tr><td rowspan="2">Shale or granite</td><td>Schist</td><td rowspan="2">Coarse-grained</td></tr>
<tr><td>Gneiss</td><td>High</td></tr>
<tr><td rowspan="3">Basalt</td><td>Greenschist</td><td>Fine-grained</td><td>Low</td></tr>
<tr><td>Amphibolite</td><td rowspan="2">Coarse-grained</td><td>Medium</td></tr>
<tr><td>Granulite</td><td>High</td></tr>
</table>

Still further metamorphism leads to a coarse-grained rock with pronounced schistosity, called a **schist.** The most obvious differences between slate, phyllite, and schist are in grain size, but the mineral assemblages change, too. At high grades of metamorphism, the minerals may segregate into bands. A high-grade rock with coarse grains and pronounced foliation, and with bands of micaceous

FOLIATION. The foliation in this rock, a gneiss, developed as a result of metamorphism. The dark-colored layers are rich in biotite and amphibole. (Courtesy Brian Skinner)

minerals segregated from bands of quartz and feldspar, is called a **gneiss** (pronounced "nice," from the old German word *gneisto,* "to sparkle").

The main minerals in basalt are olivine, pyroxene, and plagioclase, which are anhydrous (lacking water). When a basalt is subjected to metamorphism under conditions where water can enter the rock and form hydrous minerals, distinctive mineral assemblages develop. Under low grades of metamorphism, the resulting rock has pronounced foliation, like a phyllite, but it also has a distinctive green color because of the presence of chlorite (a green, micalike mineral); it is called a **greenschist.** When a greenschist is subjected to higher-grade metamorphism, chlorite is replaced by amphibole; the resulting coarse-grained rock is called an **amphibolite.** At the highest grades of metamorphism, amphibole is replaced by pyroxene, and a foliated rock called **granulite** develops.

Now let's briefly consider the classification of rocks that lack foliation (see Table 8.3). Two kinds of sedimentary rock consist entirely of a single mineral (they are "monomineralic"). One is sandstone, a clastic sedimentary rock that is made up primarily of quartz grains. The other is limestone, a chemical sedimentary rock composed primarily of calcite. Neither a pure quartz sandstone nor a limestone contains the necessary ingredients to form micas or other minerals that might impart foliation to a metamorphic rock. As a result, marble and quartzite—the metamorphic rocks derived from quartz sandstone and limestone, respectively—usually lack foliation.

Marble consists of a coarsely crystalline, interlocking network of calcite grains. During recrystallization of a limestone, the bedding planes, fossils, and other textural features of sedimentary rocks are largely obliterated. **Quartzite** is derived from sandstone by the filling in of the spaces between the original grains with silica and by recrystallization. Sometimes the ghostlike outlines of the original sedimentary grains can still be seen, even though the rock has recrystallized completely. Marbles and quartzites can result from either regional or contact metamorphism. Another kind of metamorphic rock that is commonly unfoliated is **hornfels,** a fine-grained rock that forms as a result of the contact metamorphism of shale.

Table 8.3 Classification of Unfoliated Rocks

			Metamorphism	
Original Rock	**Metamorphic Rock**	**Texture**	**Kind**	**Grade**
Limestone	Marble	Coarse-grained	Contact or regional	Medium to High
Sandstone	Quartzite	Coarse-grained	Contact or regional	Medium to High
Shale	Hornfels	Fine-grained	Contact	Medium to High

Why are quartzite and marble commonly unfoliated?

___

___

Answer: They lack the necessary mineral (and therefore chemical) constituents to crystallize into mica or another platy mineral that could impart foliation.

Test your understanding of the processes involved in metamorphism and rock deformation by taking the Self-Test.

SELF-TEST

These questions are designed to help you assess how well you have learned the concepts presented in chapter 8. The answers are given at the end.

1. A nonpermanent change in the shape or volume of a solid is called ___________ deformation.
   a. ductile
   b. plastic
   c. brittle
   d. elastic
2. Stress that is the same in all directions is called ___________ stress. Stress that is stronger in one direction than another is called ___________ stress.
   a. perpendicular; parallel
   b. hydrostatic; lithostatic
   c. uniform; differential
   d. strain; confining pressure
3. Zeolite facies metamorphism is characteristic of ___________.
   a. contact metamorphism
   b. burial metamorphism
   c. subduction zones
   d. continent-continent collisions
4. Which one of the following metamorphic rocks is not characterized by strong foliation?
   a. phyllite
   b. schist
   c. quartzite
   d. amphibolite

5. A(n) ___________ is a pair of normal faults in which the fault planes dip toward one another and the block between them has dropped down. A(n) ___________ is a pair of normal faults in which the fault planes dip away from one another and the block between them has popped up.

6. A fold in the form of an arch, with rock layers bent convex upward, is called a(n) ___________. A fold in the form of a trough, with rock layers bent concave upward, is called a(n) ___________.

7. Tension is stress that acts parallel to a surface. (T or F)

8. In a normal fault, the hanging-wall block moves down relative to the footwall block. (T or F)

9. If stress is applied very quickly, materials are more likely to behave brittlely than if the stress is applied slowly. (T or F)

10. The dividing line between processes that we call "metamorphism" and processes that we call "diagenesis" is below a depth of about 5 km, where the temperature is about 150°C. (T or F)

11. Different mineral assemblages that are formed in rocks of different composition under the same set of temperature and stress conditions are said to belong to the same ___________.

12. The process whereby the chemical composition of a rock is altered by the addition or removal of material by solution in fluids is called ___________.

13. What are the main differences between tension, compression, and shear stress?

____

____

____

____

14. What is the difference between elastic deformation and ductile (plastic) deformation?

____

____

15. What are the contour lines on a topographic map?

____

ANSWERS

1. d
2. c
3. b
4. c
5. graben; horst
6. anticline; syncline
7. F
8. T
9. T
10. T
11. metamorphic facies
12. metasomatism
13. Tension is stress that pulls or stretches rocks; it acts perpendicular to and away from a surface, and may result in a change in shape or an increase in volume of a rock. Compression is stress that squeezes rocks; it acts perpendicular to and toward a surface, shortening or decreasing the volume of a rock. Shear stress acts parallel to a surface; it causes a rock to change shape by bending, flowing, or breaking.
14. Elastic deformation is nonpermanent deformation; ductile deformation is a type of permanent deformation in which the material changes its shape by flowing or bending.
15. Lines of equal elevation.

KEY WORDS

amphibolite
anticline
asymmetrical fold
axial plane
basin
brittle deformation
burial metamorphism
compression
confining pressure
contact metamorphism
contour line
deformation
dip
dome
ductile deformation
elastic deformation
fold
fold axis
foliation
formation
geologic cross section
geologic map
gneiss
graben
granulite
greenschist

high-grade (metamorphism)
hornfels
horst
isoclinal fold
low-grade (metamorphism)
marble
metamorphic facies
metamorphism
metasomatism
migmatite
monocline
normal fault
outcrop
overturned fold
phyllite
pressure
prograde metamorphism
quartzite
recumbent fold
regional metamorphism
relief
retrograde metamorphism
reverse fault
scale
schist
schistosity
shear stress
slate
slaty cleavage
strain
strain rate
stress
strike
strike-slip fault
structural geology
symmetrical fold
syncline
tension
thrust fault
topographic map

9 The Hydrosphere and the Atmosphere

Rain added to a river that is rank
Perforce will force it overflow the bank.

—William Shakespeare

When the well's dry, we know the worth of water.

—Benjamin Franklin

Mais où sont les neiges d'antan? (But where are the snows of yesteryear?)

—François Villon

OBJECTIVES

In this chapter you will learn

- how water moves from one reservoir to another on and under the Earth's surface, and in the atmosphere;
- how we access freshwater as a resource;
- how the ocean and the atmosphere work together to determine climate;
- how both natural and anthropogenic (human-generated) processes influence global climatic change.

1 RESERVOIRS AND PROCESSES IN THE HYDROLOGIC CYCLE

Four great reservoirs make up the integrated Earth system: the lithosphere, the hydrosphere, the biosphere, and the atmosphere. The hydrosphere, as defined in chapter 2, comprises the oceans, lakes, and streams; underground water; and glaciers. However, water is present in all four reservoirs of the Earth system, not just in the hydrosphere. Water vapor is an important part of the atmosphere, the envelope of gases that surrounds the Earth. Water is a constituent of many common minerals (such as micas and clays) in the lithosphere, where it is tightly bonded in their crystal structures. And, of course, water is a fundamental component of living things in the biosphere.

The water cycle, or hydrologic cycle, describes the movement of water from one reservoir to another. Scientists who study the movement, characteristics, and distribution of water are **hydrologists** (from the Greek word *hydro,* "water"). The water cycle consists of interconnected reservoirs and pathways. The reservoirs are storage tanks where water resides for varying lengths of time; the pathways are the processes by which water is transferred among the reservoirs.

The largest reservoir consists of the world's oceans, which hold 97.5 percent of the water in the hydrosphere. Thus, most of the water moving in the hydrologic cycle is saline. This has important implications for humans because we depend on freshwater for drinking, agriculture, and industrial use. The largest reservoir of freshwater is the polar ice sheet, containing almost 74 percent of all freshwater. Groundwater accounts for almost 98.5 percent of the remaining unfrozen freshwater. A very small fraction of the water in the hydrologic cycle resides in surface freshwater bodies or in the atmosphere, and even less is stored in the biosphere.

The hydrologic cycle is powered mainly by the Sun. Heat from the Sun causes **evaporation,** the process by which water changes from liquid to vapor. Water evaporates from the oceans, from surface water bodies, from vegetation, and from land. Depending on local conditions of temperature, pressure, and humidity, some of the water vapor that enters the atmosphere will undergo **condensation**—that is, change from a vapor into a liquid or solid. In liquid or solid form it can fall as **precipitation** (rain, snow, or hail).

Rain that falls on land may evaporate, be intercepted by plants, or drain over the land, becoming **surface runoff.** It may be stored temporarily in surface water bodies such as lakes or wetlands. Or it may work its way into the ground by **infiltration,** passing through small openings and channels in the soil. Some of the water that infiltrates will eventually become part of the vast reservoir of groundwater, where it may reside for hundreds of years. Water stored in the snow and ice of glaciers may remain locked up even longer, sometimes for hundreds of thousands of years, but eventually it, too, melts or evaporates and returns to the oceans.

The processes that control the movement and distribution of water are important in everyday life. During extended periods with below-average precipitation, there are droughts. Where infiltration rates are low and precipitation is high, water accumulates at the surface and floods may occur. Water also performs environmental services, such as the dilution and removal of wastes. This sometimes causes problems; for example, rain falling on a landfill may dissolve noxious chemicals, carrying them away to contaminate a stream or groundwater supply. Moving water transports sediment, which can also cause problems; we may lose valuable topsoil, a hillslope may become unstable, or a stream channel may become clogged with sediment. Large bodies of surface water—especially the oceans—control the weather and influence the distribution of climatic zones. Even the landscapes around us have been produced by the erosional and depositional work of streams, waves, and glaciers, combined with tectonic, volcanic, and rock deformation processes. In all these ways and more, the hydrologic cycle influences our everyday lives.

What is the energy source for processes in the hydrologic cycle?

Answer: The Sun.

2 FRESHWATER ON THE SURFACE

Surface freshwater bodies—streams, rivers, lakes, ponds, and wetlands—hold only a tiny fraction of the water in the hydrosphere, but they are our most accessible source of freshwater. People throughout the world depend on rivers, streams, and lakes for drinking water, agriculture, transportation, and industrial use. Freshwater ecosystems provide habitats for plants and animals. They are also important recreation sites and sources of natural beauty.

The processes by which water collects into stream channels and flows over the land are discussed in chapters 6 and 7. Every stream is surrounded by its **drainage basin,** the total area from which water drains into the stream (Figure 9.1). In general, the greater a stream's annual discharge, the larger its drainage basin. The vast drainage basin of the Mississippi River encompasses more than 40 percent of the total area of the contiguous United States. The line that separates adjacent drainage basins is a **divide.** Divides are topographically higher than the basins they separate. On continents, great mountain chains separate streams that drain toward one side of the continent from streams that drain toward the other side. The continental divide of western North America lies along the length of the Rocky Mountains. Streams to the east ultimately drain into the Atlantic Ocean, while those to the west drain into the Pacific Ocean.

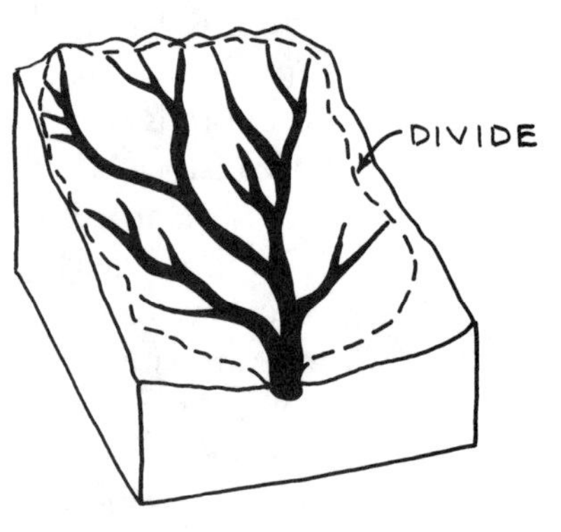

DRAINAGE BASIN
Figure 9.1

All natural streams flood from time to time, when their discharge is greater than the channel can accommodate. The geologic record shows that flooding has occurred throughout most of the Earth's history. Major floods can be disastrous events, causing both loss of life and extensive property damage. The Huang He in China, called the Yellow River because of the yellowish brown color produced by its heavy load of silt, has a long history of catastrophic floods. In 1931, a flood on the Huang He killed a staggering 3,700,000 people.

Because floods can be so damaging, much harm can be avoided by predicting them. To do this, the frequency of occurrence of past floods is plotted on a graph, producing a **flood-frequency curve** (Figure 9.2). The average time interval between floods of the same or greater magnitude is called the **recurrence interval.** A recurrence interval of 10 years means that there is a 1 in 10 (10 percent) chance that a flood of that magnitude will occur in any particular year; this is a "10-year flood." Short-term flood prediction, or forecasting, specifies the magnitude of the flood's **peak** or **crest** (its highest discharge, expressed as water level above a reference point), and the time when it will pass a particular location. Forecasting is based on real-time monitoring of storms, combined with knowledge about the topography, vegetation, and impermeable ground cover in the area. This information is used to predict the amount of surface runoff, its velocity, and its probable course, and to issue early warnings to communities that may be affected.

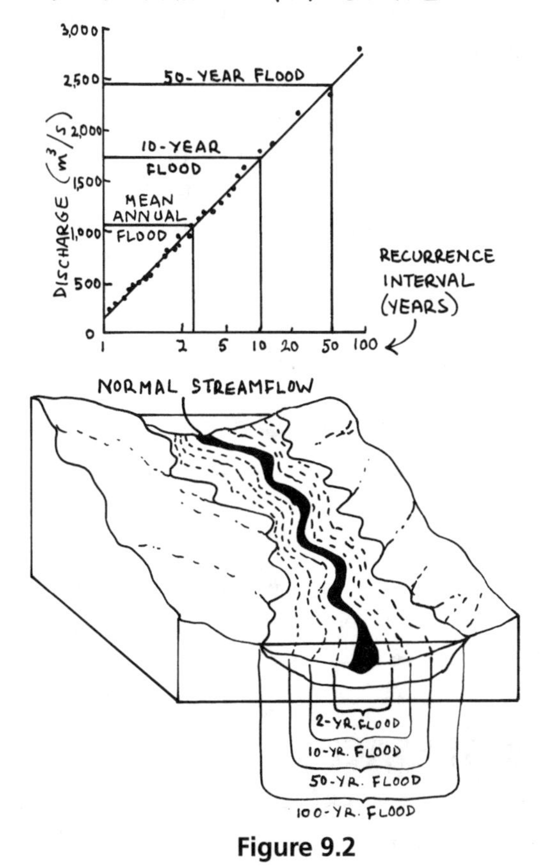

Figure 9.2

River channels are often modified or "engineered" for flood control and to increase access to floodplain lands, facilitate transportation, enhance drainage, and control erosion. The modifications usually consist of some combination of straightening, deepening, widening, clearing, or lining of the natural channel, collectively referred to as **channelization.** In the context of flood control, channelization is generally designed to increase the channel's cross-sectional area. Recall that discharge is the velocity of flow multiplied by the cross-sectional area of the channel (see chapter 6); thus, an increase in the width and depth of a channel should enable it to handle a greater discharge at a higher velocity. However, channelization is controversial because it can interfere with natural habitats and ecosystems. The aesthetic value of the river can be degraded, groundwater disrupted, and water pollution aggravated. Paradoxically, channelization may control flooding in the immediate area while contributing to more intense flooding further downstream.

Urban development can also contribute to flooding. Construction on unconsolidated, compressible sediments can lead to **subsidence** (lowering of the ground surface). The increase in impermeable ground cover associated with urbanization can add substantially to surface runoff in urban areas. Storm sewers allow runoff from paved areas to reach river channels quickly, increasing the discharge. Floods in urbanized basins often have higher peak discharges and higher total discharges and reach their peak more quickly than floods in undeveloped (natural) basins.

Lakes, ponds, and wetlands are standing surface water bodies. The scientific study of lakes and other inland bodies of water is called **limnology** (from the Greek *limne,* "lake or pond"). Lakes are generally short-lived features on the geologic time scale. They disappear by one of two processes, or a combination of both. First, lakes that have stream outlets are gradually drained as the outlets are eroded to lower levels. Second, lakes accumulate inorganic sediment carried by streams entering the lake and organic matter produced by plants within the lake. Eventually they fill up, forming boggy wetlands with little or no free water surface; this process is called **eutrophication.**

Lakes can also appear and disappear with changes in climate. In moist climates, the water level of lakes and ponds coincides with the local water table. Seepage of groundwater into the lake, combined with runoff and precipitation, maintains these water surfaces throughout the year. If the temperature increases or precipitation decreases (or both), evaporation may exceed input and the lake will shrink. In regions where climatic conditions consistently favor evaporation over precipitation, lake beds may be dry or only intermittently filled with water. Streams bring dissolved solids—salts—to these ephemeral lakes. Because evaporation removes only pure water, the salts remain behind and salinity levels may build up.

How can urbanization contribute to flooding?

____

____

____

Answer: Construction on unconsolidated, compressible sediments can lead to subsidence and urban flooding. Impermeable ground cover adds to surface runoff. Storm sewers allow runoff from paved areas to reach river channels quickly, adding to the discharge.

3 FRESHWATER UNDER THE SURFACE

Less than 1 percent of all water in the hydrologic cycle is groundwater, or subsurface water contained in spaces within bedrock and regolith. Still, the volume of groundwater is 40 times greater than that of all the water in freshwater lakes or streams. Water is present everywhere beneath the Earth's surface, even in hot deserts. At depth there is little groundwater because the pressure exerted by overlying rocks is great, and the spaces are very small.

If you were to dig a well, the hole would ordinarily pass first through a layer of moist soil and then into a zone in which the spaces in regolith or bedrock are filled mainly with air (Figure 9.3). This is the **aerated** (or **unsaturated**) **zone;** although some water may be present, it does not completely saturate the ground. Below the aerated zone is the **saturated zone,** in which all spaces are filled with water. The top surface of the saturated zone is the **water table,** the upper limit of readily usable groundwater. The water table tends to imitate the shape of the land surface: it is high beneath hills, and lower beneath valleys. Where the water table intersects the land, there will be a surface water body such as a stream or lake. If all rainfall were to cease, the water table would slowly flatten; streams, lakes, and wells would dry up as the water table fell.

Water from rainfall or snowmelt soaks into the soil by infiltration and seeps downward under the influence of gravity until it reaches the water table. This replenishment of groundwater is called **recharge.** Once in the saturated zone, groundwater tends to flow toward surface streams or lakes, where **discharge** occurs. In a discharge zone, subsurface water becomes surface water by flowing out as a spring or by joining a surface water body. The amount of time water takes to move through the ground to a discharge zone depends on the distance and rate of flow. Unlike the swift flow of rivers, which is measured in kilometers per hour, the movement of groundwater is usually measured in centimeters per day or meters per year. Water in a stream flows through an open channel, but ground-

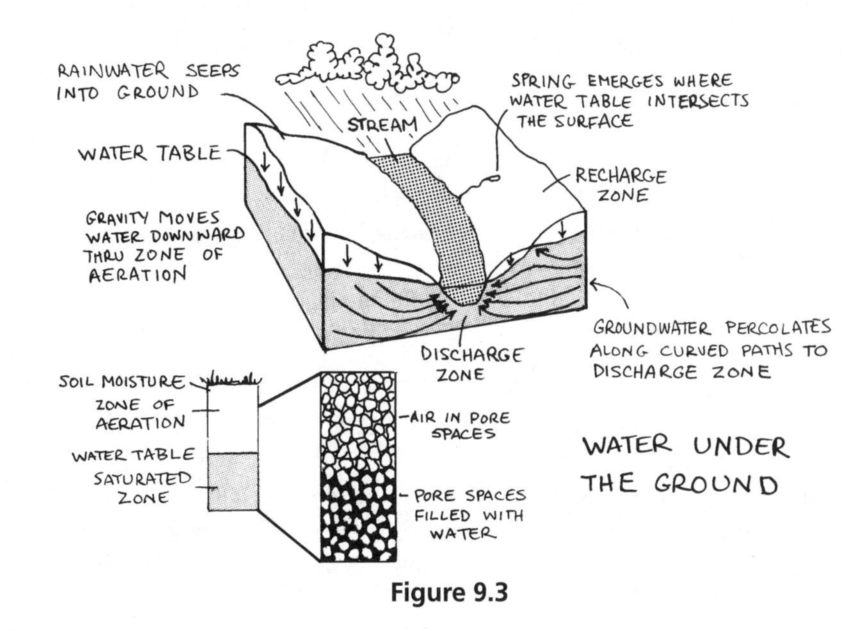

Figure 9.3

water must move by through small, constricted passages. This type of flow in the saturated zone is called **percolation.**

The rate of groundwater percolation is influenced by the nature of the rock or sediment through which the water moves, especially its porosity and permeability. **Porosity** is the percentage of the total volume of a body of rock or regolith that consists of open spaces, or pores (Figure 9.4). Porosity determines the amount of fluid a sediment or rock can hold. The porosity of sediments is affected by the sizes and shapes of the rock particles and the compactness of their arrangement. The porosity of a sedimentary rock is further affected by the extent to which the pores have been filled with cement. Unfractured plutonic igneous and metamorphic rocks, which consist of interlocking crystals, generally have lower porosities than sediments and sedimentary rocks. **Permeability** is a measure of how easily fluids can pass through a solid. A rock with low porosity is likely also to have low permeability. However, high porosity does not necessarily mean high permeability because the size of the pores and the extent to which they are interconnected also influence the ability of fluids to flow through the material. The work of groundwater in dissolving and eroding material under the ground is discussed in chapter 6.

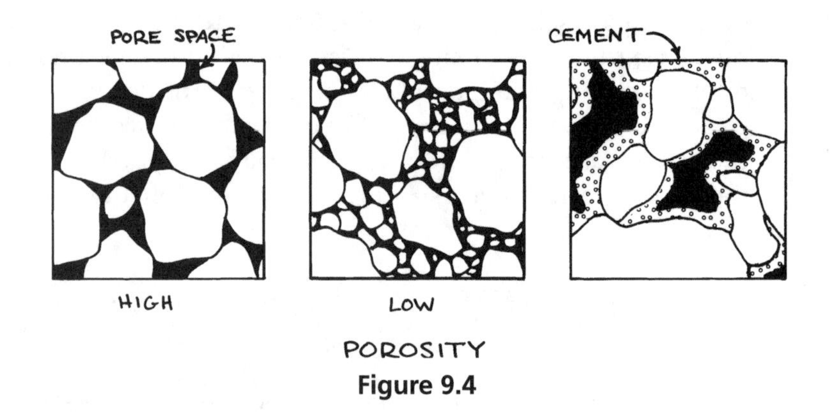

Figure 9.4

The definition given in this chapter for "discharge" is different from the definition given in chapter 6. What are the two definitions for discharge?

___

___

___

Answer: 1. The amount of water (channel width × depth × velocity of flow) passing by a point on a channel's bank during a given period of time (chapter 6). 2. The process whereby subsurface water leaves the saturated zone to become surface water (chapter 9).

4 FRESHWATER AS A RESOURCE

Without water, life on Earth would not be possible. A reliable supply of fresh water is critical for the survival and health of people and ecosystems, and for industry, agriculture, recreation, transportation, fisheries, and other activities. Globally, crop irrigation accounts for 73 percent of the demand for water, industry for 21 percent, and domestic use for 6 percent, although the proportions vary from one region to another. Water use has more than tripled since 1950. Both population growth and improved standards of living have contributed to the increase. The total amount of water withdrawn or diverted from rivers, lakes, and groundwater for human use is now about 4,340 km^3/yr (1,040 mi^3/yr), eight times the annual streamflow of the Mississippi River!

As a result of population growth and development, regions with the greatest demand for water may not have an abundant and readily available supply of surface water. For this reason, surface water is often transferred from one drainage basin to another, sometimes over long distances, via canals or pipelines. Aside from raising issues related to water rights, interbasin transfer can have negative environmental impacts. The diversion of surface water may affect the flow and salinity of the water, the amount of sediment carried by a given stream, and even the local climate. Ecosystems and water users both upstream and downstream may be affected.

Surface water bodies are highly susceptible to contamination. Many cities dump sewage directly into nearby water bodies, sometimes with little or no treatment. The accumulation of organic material—the main component of both sewage and agricultural runoff—can lead to accelerated eutrophication, in which water becomes clogged with decaying organic material. As the organic material decays, it uses up dissolved oxygen in the water, making it difficult for other life to exist there. Industrial effluents, poorly engineered landfills, and urban runoff are also important sources of surface water contamination.

Groundwater is a major source of water for human consumption, especially in dry regions where there are few streams. When we wish to find a reliable supply of groundwater, we search for an **aquifer** (Latin for "water carrier"), a body of water-saturated, porous, and permeable rock or regolith. Gravels, sands, and sandstones generally make good aquifers, for they tend to be both porous and permeable.

Aquifers are replenishable. However, if the rate of groundwater withdrawal exceeds the rate of recharge, the volume of stored water steadily decreases. It may take hundreds or even thousands of years for a depleted aquifer to be replenished. Results of excessive withdrawal include lowering of the water table; drying up of springs, streams, and wells; compaction; and subsidence. When an aquifer suffers compaction—that is, when its mineral grains collapse on one another because the pore water that held them apart has been removed—it is permanently damaged and may never be able to hold as much water as it originally held. Urban development can contribute to groundwater depletion, not only by increasing the demand for water but also by increasing the amount of impermeable ground cover. When a recharge area is covered by roads, parking lots, buildings, and sidewalks, the rate of replenishment can be substantially reduced.

Many of the pollutants that affect surface water also cause groundwater contamination. Because of its hidden nature, however, groundwater contamination can be more difficult to detect, control, and clean up. The most common sources of pollution in wells and springs are untreated sewage and agricultural chemicals. Harmful chemicals from leaking underground gasoline storage tanks, poorly designed landfills, and toxic waste facilities can also contaminate groundwater reservoirs.

Policies concerning the allocation and regulation of water use can be very controversial; water law, in general, has a complicated history. Conflicts over water rights have caused or intensified many international disputes. This continues to be true in areas where water is scarce, such as the Middle East. The application of water law to groundwater is even more complicated than for surface water because it is difficult to monitor the flow of groundwater and regulate its use. For example, if you drill a well into an aquifer underlying your property, are you entitled to withdraw as much water as you need from that well? Should you be permitted to withdraw the water and sell it elsewhere? What happens if withdrawing the groundwater depletes the aquifer and your neighbor's well runs dry? Similar problems arise when a landowner's activities cause an aquifer to become contaminated; in some areas groundwater contamination is not considered to be a problem unless the contaminant migrates underground and crosses a property boundary.

What is interbasin transfer?

Answer: Diversion of water from one drainage basin to another, sometimes over long distances, via canals or pipelines.

5 THE WORLD OCEAN

Seawater covers 70.8 percent of the surface of the Earth. The Earth's oldest rocks include sedimentary strata that were deposited by water. Therefore, we are sure that as far back as we can see, about 4.0 billion years, the Earth has had liquid water on its surface. We can be reasonably certain, then, that oceans were created sometime between 4.56 billion years ago, when the Earth formed, and 4.0 billion years ago, when the oldest known sedimentary rock was made. Where the water in the oceans came from is still an open question; it most likely condensed from steam produced by primordial volcanic eruptions.

Most of the water on our planet is contained in three huge interconnected basins—the Pacific, Atlantic, and Indian Oceans. (The Arctic Ocean is an extension of the Atlantic.) All three are connected with the Southern Ocean, the body of water that encircles Antarctica. Collectively, these four vast interconnected bodies of water, together with about 20 smaller seas, make up the "world ocean." The land that constitutes the rest of the Earth's surface is unevenly distributed. A view of the globe from directly above Russia shows mostly land; a view from directly above New Zealand shows mostly water. The uneven distribution of land and water plays an important role in determining the paths along which water circulates in the open ocean.

The composition of seawater is expressed as **salinity,** or saltiness. The salinity of seawater ranges between 33‰ and 37‰ (‰ = per mil = parts per thousand). The main elements that contribute to this salinity are ions of sodium (Na^+) and chlorine (Cl^-). When seawater evaporates, it is not surprising that most of the dissolved matter is precipitated as common salt (NaCl, the mineral halite). Seawater contains most of the other natural elements as well, but they are present in very low concentrations. The ions in seawater come mainly from the weathering of minerals. As rocks of the crust interact with the atmosphere and rainwater during weathering, chemical elements are leached out and become part of the dissolved load in stream water flowing to the sea. Other sources of ions are volcanic eruptions (on land and underwater) and dust from deserts blown out to sea.

The salinity of seawater is controlled by the interplay between several processes.

1. Evaporation removes freshwater and leaves the remaining water saltier.
2. Precipitation adds freshwater and makes seawater less salty.
3. Inflow of freshwater from rivers makes seawater less salty (even though river water carries dissolved ions, it is still less salty than seawater).
4. Freezing excludes salts from the ice (that is, ice crystals are purer than the salt water they crystallize from), leaving the unfrozen seawater saltier.

The salinity and temperature of seawater control its density. High salinity means high density; high temperature means low density. Ocean scientists (**oceanographers**) define three major density layers or zones in the ocean. The differences between these layers are caused by both temperature and salinity changes, with temperature as the major control. The **surface zone,** typically extending to a depth of 100 m (328 ft), consists of water that is relatively warm and therefore has a low density. Below the surface zone, to a depth of 500 m (1,640 ft), the water temperature decreases rapidly with depth. This is called the **thermocline.** Below the thermocline is the **deep zone,** which contains the bulk of the ocean's volume and in which temperature is low, about 2°C (about 36°F).

The salinity and temperature of ocean water drive deep-ocean currents. Seawater near Antarctica and in the Arctic Ocean is very cold; it is also quite saline because of the formation of sea ice. Low temperatures and high salinity mean high density. In polar regions, cold, salty, dense water sinks, setting into motion a deep global oceanic circulation system that has a fundamental influence on climate. Surface ocean currents are broad, slow drifts of water driven by winds. Air that flows across the sea drags the water forward, creating a current of water as broad as the current of air but confined to the surface zone, only 50 to 100 m (about 165 to 325 ft) deep. The surface and deep ocean currents are the main mixing forces in the ocean.

What are the four major processes that control the salinity of seawater?

Answer: 1. Evaporation removes freshwater, leaving seawater saltier.
2. Precipitation adds freshwater, making seawater less salty.
3. Inflow of freshwater from rivers makes seawater less salty.
4. Freezing excludes salts, leaving unfrozen seawater saltier.

6 WHERE THE OCEAN MEETS THE LAND

A majority of the world's population lives within 100 km of the ocean. This reflects our dependence on the oceans and the economic benefits afforded by acccss to ocean resources. However, the concentration of large numbers of people in coastal areas means that the coastal environment must absorb the impacts of a wide range of human activities. It is also worth noting that human vulnerability to hazards can be particularly high in coastal zones.

Visit a coastal zone on two occasions a year apart and you will see changes. Sand dunes may have shifted. Sand may have built up behind barriers or eroded away. Steep sections of coastline may have collapsed. Channels may have broken through from the sea to lagoons on the landward side, where no channels were before.

Most of the changes we see happening along the shoreline result from the action of tides and waves. **Tides** are the cycles of regular rise and fall of the level of water in oceans. The gravitational attraction of the Moon (and, to a much lesser extent because of its great distance, the Sun) causes ocean water to bulge outward on the side of the Earth nearest to the Moon. On the opposite side of the Earth, inertia created by the Earth's rotation causes ocean water to bulge outward in the opposite direction. The result is two oceanic tidal bulges, one positioned on either side of the Earth (Figure 9.5). While the Earth itself rotates, these tidal bulges remain essentially stationary. Any given coastline will move westward through both tidal bulges each day. Every time a landmass encounters a tidal bulge, the water

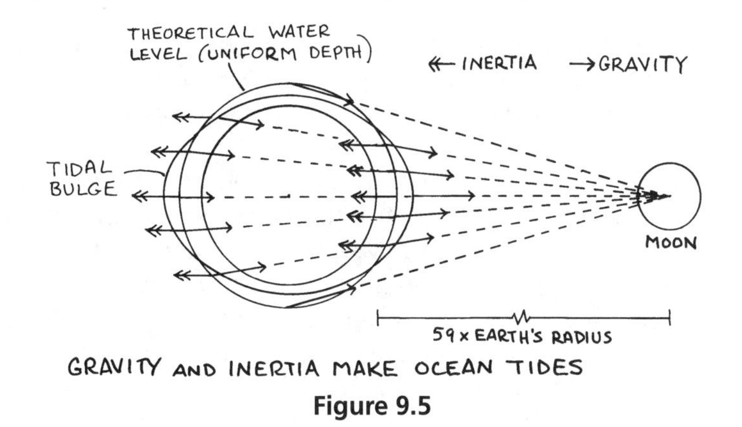

Figure 9.5

level along the coast rises. As the Earth rotates, the coast passes through the highest point of the tidal bulge (high tide) and then the water level begins to fall. There are two high tides and two low tides each day, because any given coast will necessarily pass through both tidal bulges during every complete rotation of the Earth.

The other major interaction between ocean water and coastal land comes from the action of waves. Waves, like surface currents, are driven by wind. The size of a wave depends on how fast, how far, and how long the wind blows across the water surface.

When you visit the shoreline, look carefully at the waves coming into the beach. You can usually see them arriving from two or more directions. Each set of waves originated from a storm-related wind far beyond your line of sight.

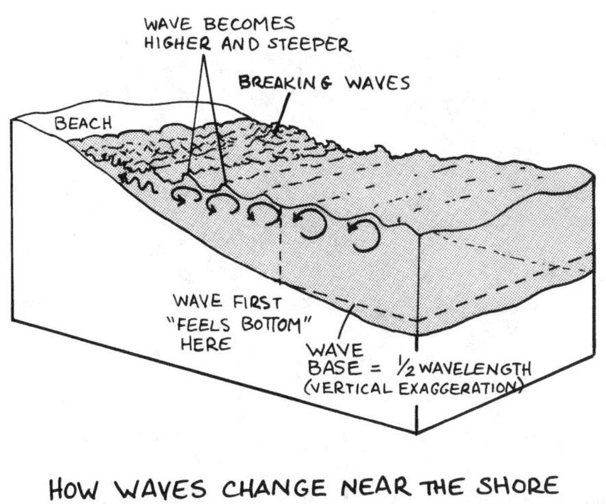

Figure 9.6

In deep water, each small parcel of water in a wave moves in a loop, returning very nearly to its former position as the wave passes (Figure 9.6). The distance between successive wave crests is the **wavelength,** denoted λ (the Greek letter lambda). Downward from the surface of the water, a progressive loss of energy occurs; this is expressed as a decrease in the diameter of the looplike motion of the water parcels. Eventually, at a depth of about $\lambda/2$ (half of one wavelength), the motion of the water becomes negligible. This lower limit of wave movement is called the **wave base;** it is also the lower limit of the erosive capability of a wave.

As a wave approaches the shore, it undergoes a transformation. Landward of depth $\frac{L}{2}$, the circular wave motion is influenced by the increasingly shallow seafloor, which restricts vertical movement. With decreasing depth, the circular orbits become progressively flatter until the movement of water is limited to a back-and-forth motion. As the shallow seafloor interferes with wave motion and distorts the wave's shape, the wave height increases and the wavelength decreases. Now the front of the wave is in the shallower water than the rear and is also steeper than the rear. Eventually the front becomes too steep to support the advancing wave. As the rear part continues to move forward, the wave collapses, or breaks.

When a wave breaks, the motion of the water becomes turbulent. Such "broken water," or **surf,** is found between the line of breakers and the shore. Surf is a powerful erosional agent because it possesses most of the original energy of each wave that created it. Wave erosion takes place not only at sea level but also below and—especially during storms—above sea level. In the surf zone, rock particles transported by waves are worn down, becoming smoother, rounder, and smaller. Through continuous rubbing and grinding with these particles, the surf wears down and deepens the bottom and eats into the land.

Most waves hit the shore at an oblique angle. This sets up a **longshore current,** flowing parallel to the shoreline within the surf zone (Figure 9.7). Surf erodes sediment, and the longshore current moves the sediment along the beach. Meanwhile, on the beach itself, each incoming wave travels at an oblique angle up the beach until gravity pulls the water straight down the slope of the beach. Successive zigzag movements gradually transport sediment along the shore, a process known

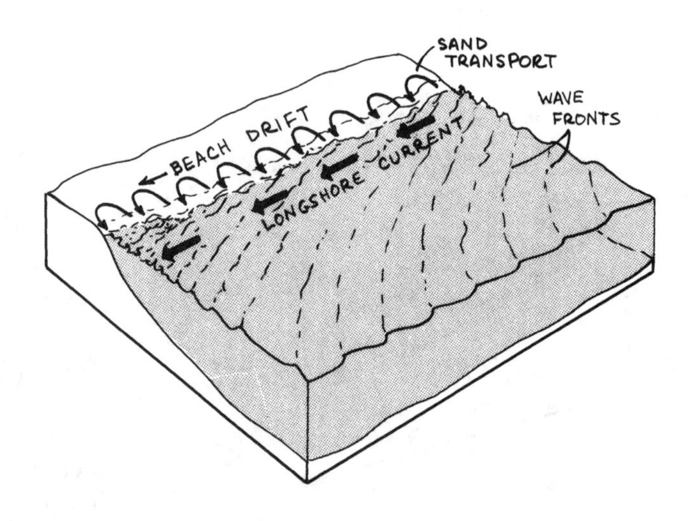

LONGSHORE CURRENT AND BEACH DRIFT

Figure 9.7

LONGSHORE CURRENT. Groins (barriers) along Westhampton Beach, Long Island, New York, were built to prevent sand from being moved along the shore by the longshore current. The view is looking eastward. (Courtesy U.S. Geological Survey)

as **beach drift.** Through these mechanisms, waves erode sediment and transport it along the shoreline, eventually depositing it further downcurrent or offshore.

If a wave has a wavelength $\lambda = 400$ m (1,300 ft), then what is the depth of its wave base?

Answer: The wave base = $\lambda/2 = 400/2 = 200$ m (650 ft).

7 THE SHAPE OF THE SHORELINE

The end result of the constant interplay between erosive and depositional forces along coastlines is a wide variety of shorelines and coastal landforms. Coastal configurations depend on the active geologic processes at work, the structure and erodability of coastal rocks, and the length of time these processes have operated. Sea level changes can also influence coastal features. Many coastal and offshore landforms are relics of times when sea level was higher or lower than it is now.

The most common type of coast, comprising about 80 percent of ocean coasts worldwide, is a rocky or cliffed coast. A **wave-cut cliff** is a coastal cliff cut by wave action at the base of a rocky coast. As the upper part of the cliff is undermined, it collapses and the resulting debris is redistributed by waves. Below a wave-cut cliff you can often find a **wave-cut bench** (or **terrace**), a platform that has been cut into the bedrock by surf action. Cliffed shorelines are susceptible to frequent landslides and rock falls as erosion eats away at the base of the cliff.

A beach consists of wave-washed sediment along a coast. A landform commonly associated with beaches is the **barrier island,** a long, narrow, sandy island lying offshore and parallel to the coast. A barrier island consists of one or more ridges of sand dunes associated with successive shorelines in a region of rising sea level. During major storms, surf washes across low places on barrier islands and erodes them, cutting inlets that may remain open permanently. In this way the length and shape of barrier islands are always changing. Barrier islands are found along most of the world's lowland coasts on trailing continental plate margins. The Atlantic Coast of the United States consists mainly of a series of barrier beaches.

Many of the world's tropical coastlines consist of limestone **reefs** built by vast colonies of tiny organisms, principally corals, that secrete calcium carbonate. Reefs are built very slowly over thousands of years. Each of the tiny coral animals deposits a protective layer of calcareous material; the layers eventually combine to form a complex reef structure. Reefs are highly productive ecosystems inhabited by a diversity of marine life-forms. They also provide physical barriers that dissipate the force of high-energy waves, protecting ports, lagoons, and beaches that lie behind them. They are an important aesthetic and economic resource. Corals

WAVE-CUT TERRACES. Wave-cut terraces in Mallagh Landing, San Luis, California, were elevated by tectonic movements. The upper terrace is now 30 m (100 ft) above sea level; the middle one, where the houses are, is 18 m (60 ft) above sea level; and the lowest terrace, where the steps and boat landing are, is 3 m (10 ft) above sea level. (Courtesy U.S. Geological Survey)

require shallow, clear water in which the temperature remains above 18°C (65°F). Because of their very specific water temperature and light-level requirements, coral reefs are particularly susceptible to damage from human activities as well as from natural causes such as tropical storms.

What are barrier islands, and in what type of tectonic environment are they found?

Answer: Long, narrow, sandy islands lying offshore and parallel to the coast; they are found along lowland coasts on trailing continental plate margins.

8 THE ATMOSPHERE

An atmosphere is an envelope of gases that surrounds a planet or a natural satellite (see chapter 2). **Air** is the invisible, odorless mixture of gases and suspended particles that surrounds one special planet, Earth. In other words, air is what Earth's atmosphere is made of. The composition of air varies from place to place and from time to time in the same place, mainly because of variations in aerosols and water vapor. **Aerosols** are liquid droplets or solid particles that are so small they remain suspended in the air. Water droplets in fog are liquid aerosols. Solid aerosols include tiny ice crystals, smoke particles from fires, sea-salt crystals from ocean spray, dust stirred by winds, and some volcanic emissions. The amount of water vapor in the air, its **humidity,** is also quite variable. On a hot, humid day in the tropics, as much as 4.0 percent of the air by volume may be water vapor. On a crisp, cold day, less than 0.3 percent water vapor may be present.

Because the water vapor and aerosol contents of air vary, the composition of air is usually reported as if the air were dry (free of water vapor) and aerosol-free. Once these two highly variable components have been removed, the relative proportions of the remaining gases are almost constant. Three gases—nitrogen, oxygen, and argon—make up 99.96 percent of dry air by volume (Figure 9.8). Other gases are present in very small quantities.

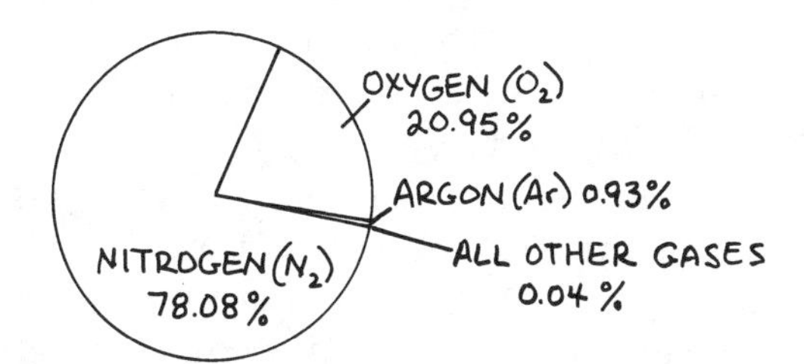

Figure 9.8

The atmosphere is composed of four layers with distinct temperature profiles (as shown in Figure 9.9), each separated by thermal boundaries called **pauses.** From the bottom up, the layers are:

1. The **troposphere,** from the ground surface up to about 15 km (9 mi)
2. The **stratosphere,** up to 50 km (30 mi)
3. The **mesosphere,** up to 90 km (56 mi); and
4. The **thermosphere,** up to 700 km (more than 430 mi)

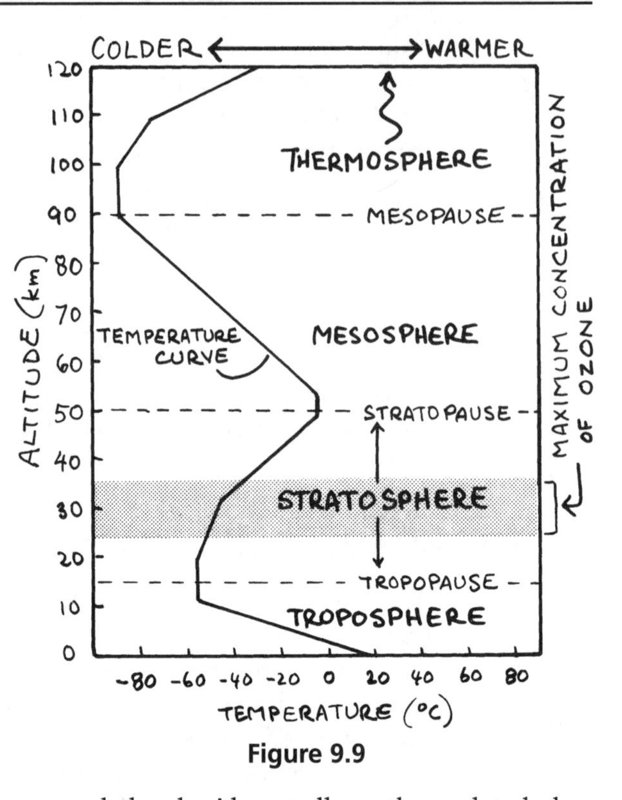

Figure 9.9

We humans live at the bottom of the troposphere. The troposphere contains 90 percent of the actual mass (that is, the matter) of the atmosphere, including virtually all the water vapor and clouds. Almost all weather-related phenomena originate in the troposphere. Although little mixing occurs between the troposphere and the stratosphere, the troposphere itself is dynamic, constantly moving, and thoroughly mixed by winds.

The troposphere contains heat-absorbing gases (notably H_2O and CO_2) that are responsible for warming the surface of the Earth. Incoming solar energy is mostly short-wavelength ultraviolet radiation. Through interactions with the atmosphere, the ocean, the land, and the biosphere, the short-wavelength radiation is changed into longer-wavelength infrared radiation—heat. In the form of heat, it is reradiated from the Earth's surface back into outer space (Figure 9.10). Some of the outgoing energy encounters **radiatively active gases** (also called **greenhouse gases**) that absorb heat. Some heat is thus retained in the lower atmosphere, causing the temperature of the Earth's surface to rise. This natural atmospheric process is the **greenhouse effect.** Without the greenhouse effect, the surface of the Earth would be cold and inhospitable to life as we know it.

The stratosphere contains another 19 percent of the atmosphere's total mass. (This leaves about 1 percent of the atmosphere's mass to reside in the upper 250 km, or 150 mi, of the atmosphere.) The stratosphere plays an important role in the global distribution of materials in the atmosphere. If pollutants, dust, or volcanic emissions are injected into the atmosphere high enough to reach the stratosphere, they may be circulated globally within a matter of days or weeks. The stratosphere

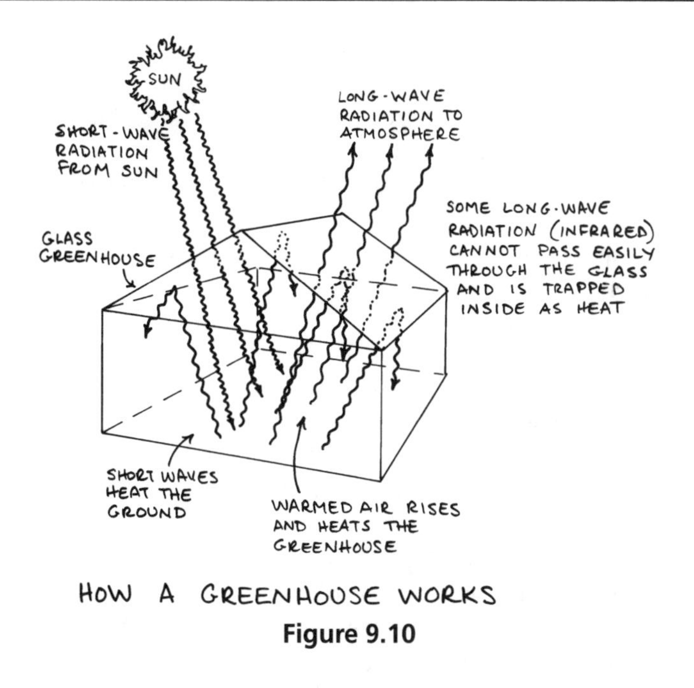

Figure 9.10

also contains a concentration of the gaseous chemical ozone (O_3). Ozone protects life by absorbing harmful short-wavelength ultraviolet solar radiation, preventing it from reaching the surface. This concentration of O_3 is called the **ozone layer.** Recently there has been much concern about the breakdown of stratospheric ozone caused by anthropogenic chemical pollutants.

What is the proper scientific term for "greenhouse gases"?

Answer: Radiatively active gases.

9 ATMOSPHERIC CIRCULATION

Because the Earth is a sphere, the Sun's rays do not reach the surface in equal amounts everywhere. Where the Sun is exactly overhead, the incoming rays are perpendicular to the surface. But everywhere else, because of the Earth's curvature, the surface is at an angle to the incoming rays. Therefore, less energy reaches each square meter of surface area. Atmospheric circulation acts to smooth out the temperature differences that result from the unequal heating of the surface, by transporting heat and moisture from one part of the globe to another.

To understand how atmospheric circulation works, remember that most materials expand when heated, becoming less dense. In contrast, when materials cool, they typically become denser. Thus, heated air near the equator expands, becomes lighter, and rises. Near the top of the troposphere it spreads outward in the direction of both poles. As the upper air travels away from the equator, it gradually cools, becomes heavier, and sinks. On reaching the Earth's surface, this cool, descending air flows back toward the equator, warms up, and rises. This movement—rising of hot material, lateral flow, and sinking of cooler, denser material—forms a convection cell.

Do you recall learning about another context in which convection is important in the Earth system?

____

Answer: Plate tectonics and mantle convection, in chapter 1.

If the Earth did not rotate, the convection currents of air would rise at the equator, flow from the equator to the poles at high altitude, then cool and sink, flowing back to the equator along the surface. But the Earth does rotate, and the rotation complicates the simple convection currents in the atmosphere (and in the ocean). The **Coriolis effect,** named after the nineteenth-century French mathematician who first analyzed it, causes anything that moves freely with respect to the rotating Earth to veer to the right in the Northern Hemisphere and to the left in the Southern Hemisphere. Both flowing water and flowing air respond to the Earth's rotation. Thus, the global patterns of both ocean currents and wind systems are influenced by the Coriolis effect.

In the atmosphere, the Coriolis effect breaks up the flow of air between the equator and the poles into belts (Figure 9.11). For example, a large belt or cell of circulating air lies between the equator (0° latitude) and about 30° latitude in both the Northern and Southern Hemispheres. Warm air rises at the equator. It begins to move toward the poles, but is deflected by the Coriolis effect. By the time it reaches a latitude of 30°, the high-altitude air mass has cooled. It meets another mass of cool air flowing from the poles toward the equator, and both air masses descend. Some of the descending air flows back toward the equator, where it will rise again. The low-latitude cells (from 0° to 30° N and S) are called **Hadley cells.** In each hemisphere, a second wind system or cell of circulating air lies poleward of the Hadley cells. These are called **Ferrel cells.** A third set of circulating air cells, called **polar cells,** lies over the polar regions. In each polar cell, cold, dry, upper air descends near the pole and moves toward the equator. As this air slowly warms, it rises along the polar front (the zone where the polar cells and the Ferrel cells meet) and returns toward the pole.

Where is the ozone layer, and what does it do?

____

____

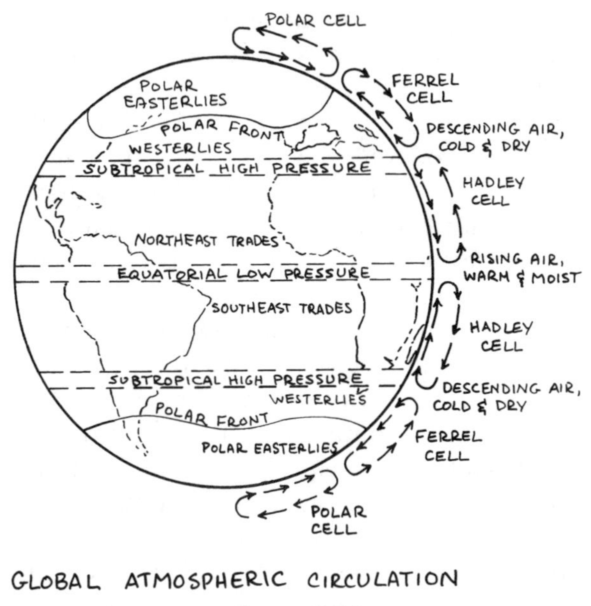

GLOBAL ATMOSPHERIC CIRCULATION

Figure 9.11

Answer: The ozone layer is found in the stratosphere; it prevents harmful short-wavelength radiation from reaching the surface of the Earth.

10 CLIMATE AND WEATHER

The ocean and the atmosphere are inextricably linked; they are complementary parts of one huge, complex, dynamic system. The ocean plays a critical role in climate and weather systems, particularly in regulating the temperature and humidity of the lower part of the atmosphere. Atmospheric circulation, in turn, drives ocean waves and currents and transfers heat. Global patterns of airflow are influenced by the uneven heating of the Earth's surface, the Coriolis effect, the distribution of land and sea, and the topography of the land.

Before proceeding, we should clarify the differences between weather and climate. **Weather** refers to local conditions in the atmosphere at any given time, usually described in terms of temperature, pressure, humidity, cloud cover, precipitation, and wind velocity. When weather patterns for a given region are averaged over a significant period, the term **climate** is used. The patterns and processes that we think of as weather—wind, rain, snow, sunshine, storms, even floods and droughts—are just temporary local variations when viewed against the more stable, longer-term background of global climate. The climate classification for a given area takes into account the average, or "normal," weather conditions, as well as the types of weather extremes that occur, their likelihood, and their frequency.

The interactions among air, land, and water that create weather are not only complex, but also highly sensitive to changing conditions. **Meteorologists** (scientists who study weather) sometimes characterize this sensitivity by saying that a change as small as the draft created by a butterfly's wing can become magnified through a network of feedback systems, ultimately developing into a wind or even a cyclone. They even have a name for this phenomenon: the "butterfly effect." The variability, complexity, and extreme sensitivity of the atmosphere-ocean system make weather prediction a very tricky business.

Some weather that we think of as "extreme" is actually characteristic of the climate zone in which it occurs and can be explained by examining the movement of air masses in the region. An example is the torrential rains and violent storms typical of the Asian **monsoon.** For half a year during the winter months, the wind blows to the south across India from the high, cold plateau of central Asia. Here, just south of the equator, the Sun is overhead in the winter and the warm air rises. During the summer, the pattern is reversed. With the Sun overhead in the Northern Hemisphere, the landmass of Asia heats up and is beset by cyclones. Warm, moisture-laden winds then blow in the opposite direction, from the Indian Ocean onto the land. Summer is a time of hot, humid weather and heavy rains. The reversing winds that characterize the monsoon take their name from the Arabic word *mausim,* which means "change."

Climatologists (scientists who study climate) use several different classification schemes to describe and categorize the Earth's climate zones. One of the most widely used is the Köppen-Geiger climate system, first devised in 1918. Vladimir Köppen defined the boundaries of climate zones so that they coincided closely with the boundaries between major vegetation types. Like other classification systems, the Köppen-Geiger system defines climate zones on the basis of temperature and precipitation, two climatic factors with a major influence on vegetation. The six basic types of climate zones are tropical, dry, temperate-humid, cold-humid, polar, and highland.

What is the difference between a meteorologist and a climatologist?

____

____

Answer: A meteorologist studies weather. A climatologist studies climate (weather conditions averaged out over a long period).

11 CLIMATIC CHANGE THROUGH GEOLOGIC TIME

Periods during which the average temperature at the Earth's surface drops by several degrees and stays low long enough for existing ice sheets to grow larger (and new

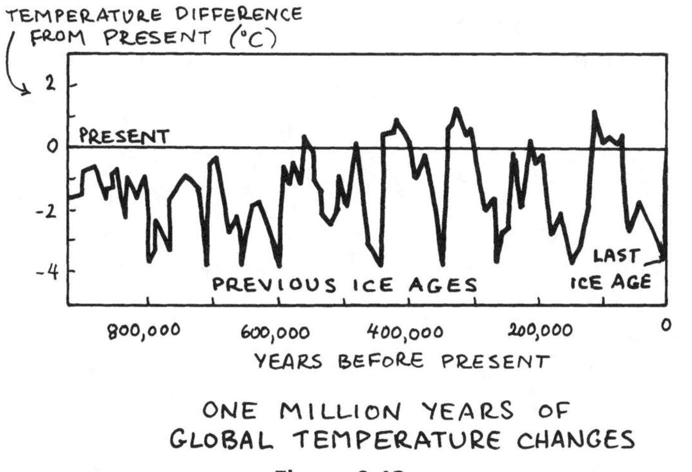

Figure 9.12

ones to form) are called **glaciations** (also called ice ages, glacial periods, glacial stages, or glacial epochs). Periods between glaciations, when the ice sheets retreat and sea levels rise, are called **interglacials** (also called interglacial stages or interglacial periods). During the past 1.6 million years—the Pleistocene and Holocene epochs—the Earth has experienced more than 20 glacial-interglacial cycles. The timing of these cycles has varied, with extreme temperature minima (low points) occurring roughly every 100,000 years over the past million years, and every 20,000 to 40,000 years before that (Figure 9.12). Glaciation has been especially widespread during the Pleistocene epoch, but it is not a new phenomenon. The rock record contains evidence of glacial ages that occurred as long ago as 2.3 billion years.

How do geologists know what the Earth's climate and surface temperature were like 20,000 or 100,000 or even 100 million years ago? Historical records of temperature and other aspects of weather have only been kept on a regular basis since the mid-1800s. Through **paleoclimatology,** the study of ancient climates, scientists employ a variety of techniques to extend these records.

For example, paleontologists infer past climates from assemblages of fossil plants and animals. Fossilized pollen spores have been particularly useful in reconstructing past climatic changes. Sedimentologists and stratigraphers can infer many things about past climates from the nature of the sediments and rocks they study. The sedimentary rock record reveals the past distribution of sediments and the mechanisms of weathering, transport, and deposition. Sediments and sedimentary rocks can also preserve a record of changes in the chemical composition of water and air. Ancient soil horizons, called paleosols, can provide a great deal of information about climate and weather in ancient environments. Geochemists can learn about the environments in which weathering has occurred by examining the minerals in paleosols; these enable them to determine the chemical compositions of the air and water at the time that the soil was formed.

Former ground surface temperatures can be determined by studying variations in temperature profiles preserved in sediment layers. Layered lake-bottom sediments and corals can reveal fine seasonal variations. The amount of dust in the atmosphere in earlier periods can be deduced from the distribution of eolian sediments. Isotope geochemists can determine past temperatures from studies of sediments and core samples from polar ice. Scientists have even obtained samples of ancient air, up to 170,000 years old, trapped in bubbles in glacial ice. Chemical tests on this air can reveal its composition, and tests on the ice itself can reveal whether global temperatures were relatively warm or cool at the time the ice was deposited.

The most recent glaciation began about 30,000 years ago. Geologists have been able to determine the timing, extent, and nature of the recent Ice Age (and earlier glaciations as well) from a variety of evidence. For example, rocks that were scratched and grooved by glaciers reveal the direction in which the ice was moving. Glacial landforms, especially moraines, reveal the geographic extent of land ice sheets (see chapter 6 for a review of glacier terminology). The radiocarbon ages of trees that were felled by advancing ice tell geologists when the ice arrived in a given region. Great thicknesses of windblown loess were deposited just south of the ice limits during glacial times. These deposits, which contain fossil plants and animals characteristic of cold, dry weather, reveal that the Ice Age climate was both colder and dustier than today's climate.

The Pleistocene glaciation was a time when great woolly mammoths, mastodons, longhorn bison, and saber-toothed tigers roamed North America. Early humans migrated into North America from Asia, walking across exposed continental shelf in today's Bering Strait in Alaska. (The land was exposed during the Ice Age because a large volume of water was locked up in glacial ice, causing sea levels to drop globally.) The early humans used stone and wooden tools to hunt the large mammals, perhaps driving some of them to extinction. Huge floating ice sheets like those found in present-day Arctic and Antarctic seas occupied large areas of the Atlantic Ocean.

About 10,000 years ago, the Earth emerged from the Ice Age. We have now passed the time of maximum warmth in the glacial-interglacial cycle; temperatures peaked in a warm period about 6,000 to 7,000 years ago, an interval known as the Holocene Optimum. Since then temperatures have been gradually cooling, with some distinctly cooler fluctuations (such as the Little Ice Age, a generally cool period that lasted from about A.D. 1300 to about A.D. 1900).

When did the most recent Pleistocene Ice Age begin, and when did it end?

Answer: It began 30,000 years ago and ended 10,000 years ago.

12 CAUSES OF CLIMATIC CHANGE

What factors cause the Earth's climate to change? The search for an answer has been difficult because the climate system is very complicated, with many interacting subsystems. Climate changes on several different time scales, ranging from decades to many millions of years. And modern impacts of human activities on the climate system are making it increasingly difficult to separate natural from anthropogenic (human-generated) influences. By reconstructing past climates, geologists can determine the range and time scales of climatic variations. Then they can use the data to test the accuracy of computer models that simulate both past and future climatic change. We know that the Earth's climate *will* change, for the geologic record of climatic change is clear. What we lack is a clear understanding of *how* it will change and at what rate.

Several mechanisms cause natural climatic changes. They involve the atmosphere, the lithosphere, the ocean, and the biosphere, all interacting in complex ways. For example, geographic changes resulting from tectonism—the shifting of continents, the uplift of continental crust and creation of large mountain chains, and the opening or closing of ocean basins—have a significant impact on oceanic and atmospheric circulation and, therefore, on global climate. Even the Earth's internal processes can affect climate. For example, large, explosive volcanic eruptions sometimes produce vast quantities of dust and tiny aerosol droplets of sulfuric acid, both of which scatter the Sun's rays and cause global cooling.

The Earth's orbit and rotation also play an important role in controlling the timing and cyclicity of climatic variations (Figure 9.13). The eccentricity (departure from circularity) of the Earth's orbit, the tilt of the planet's axis of rotation, and the precession (wobbling) of the axis all have an impact on **insolation,** the amount of solar radiation reaching the Earth's surface at any given time. These movements all vary on different time scales. Periodic changes in climate caused by fluctuations in insolation that result from variations in the Earth's orbital and rotational characteristics are called **Milankovitch cycles,** after a Yugoslavian mathematician who studied this phenomenon.

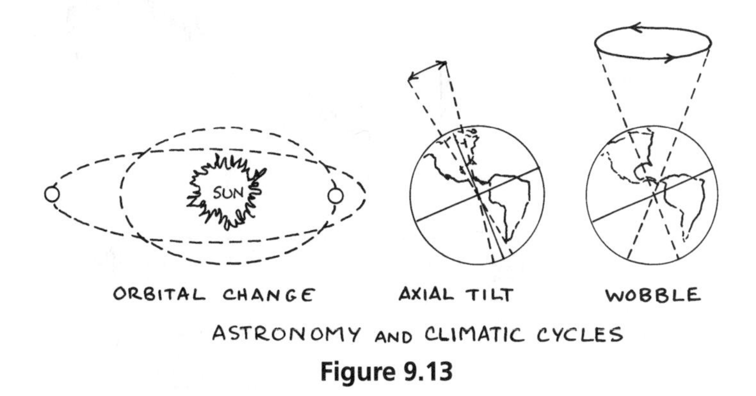

Figure 9.13

All of these influences on climate are tempered by the moderating action of the Earth's atmosphere through the greenhouse effect. Water vapor (H_2O) in the atmosphere is the most important natural greenhouse gas, but naturally occurring minor gases such as carbon dioxide (CO_2) and methane (CH_4) also store heat, providing significant warming. During the past decade or so, the possibility of accelerated global warming and the role of anthropogenic emissions have been subjects of scientific and political debates. We know that greenhouse gases are emitted into the atmosphere as a result of human activities (especially from the burning of fossil fuels). The rate of emission has been increasing dramatically since the Industrial Revolution, causing changes in the chemistry of the atmosphere. What are the effects of such changes on the global climate system?

The last few years of the 1900s were some of the warmest on record in North America. Then, in the summer of 2000, a startling and disturbing discovery was made: the ice cap that formerly covered the North Pole year-round had begun to melt. Do these observations mean that the world has inevitably started into a long-term warming trend caused by human-generated atmospheric changes? It is impossible to be certain, but the observations do seem to lend credence to the predictions of climatologists. Historical records of surface temperature have been used to estimate annual average global temperatures over the past 100 years (Figure 9.14). From these estimates it appears that global temperatures have increased by 0.5 to 0.6°C (0.9 to 1.1°F). But it is extremely difficult to determine average surface temperatures conclusively on a time scale as short as this, particularly because the existing data from the early part of the past 100 years are insufficient for the task.

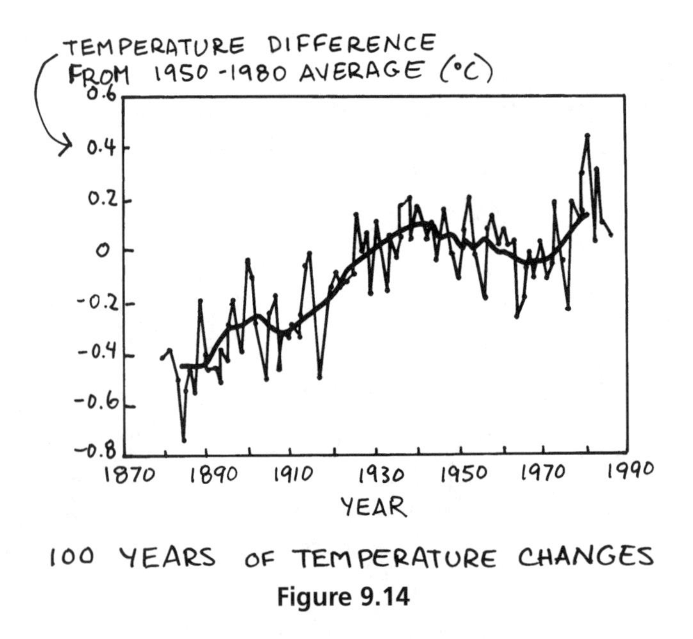

Figure 9.14

Will the warming trend continue? Is it a long-term trend? How hot will the surface temperature of the Earth become? How much of the change is natural, and how much is caused by human actions? And if the global climate does become warmer, what effects will it have on humans and ecosystems? These questions have not yet been answered conclusively. The Earth's climate system is very complex, and some parts of it are still poorly understood. For example, clouds play a dual role: they warm the surface by trapping infrared radiation, but they also cool the surface by reflecting incoming solar radiation. If the global temperature gets warmer, the rate of evaporation of ocean water will increase, perhaps causing the amount of cloud cover to increase. Will a change in cloud cover have an overall warming effect or a cooling effect? No one knows for certain. Geologists, climatologists, and other Earth scientists are working hard to understand how different parts of the Earth system work together to determine global climate, and how the system can be affected by human actions.

What is a Milankovitch cycle?

Answer: Periodic variations in climate caused by fluctuations in insolation due to changes in the Earth's orbital and rotational characteristics.

The Self-Test will help you assess your understanding and retention of the material presented in this chapter. Don't forget to test yourself on the vocabulary words as well!

SELF-TEST

These questions are designed to help you assess how well you have learned the concepts presented in chapter 9. The answers are given at the end.

1. Porosity in sediments and rocks is defined as ___________.
   a. interconnections among the pores between grains
   b. the proportion of pore space to total volume in a rock or sediment
   c. the ability to absorb water
   d. the degree of lithification

2. At a water depth of about ___________, wave motion becomes negligible.
   a. 1 m
   b. 400 m
   c. $\lambda/2$
   d. 1 km

3. The __________ is where most of the Earth's weather is generated.
   a. troposphere
   b. stratosphere
   c. mesosphere
   d. thermosphere

4. The Coriolis effect is caused by __________.
   a. unequal solar heating of the Earth's surface
   b. tidal interactions among the Earth, the Moon, and the Sun
   c. the Earth's rotation
   d. All of the above are true.

5. The water table marks the boundary between the __________ and the __________.

6. A(n) __________ is the total area that supplies water to the stream system that drains it.

7. The technical name for the flow of groundwater in the saturated zone is __________.

8. Approximately 40 percent of the water in the hydrosphere is groundwater. (T or F)

9. There have been at least 20 glacial-interglacial periods during the Pleistocene epoch. (T or F)

10. The most recent glaciation began about 100,000 years ago. (T or F)

11. How does excess organic material (such as from sewage or fertilizer) cause problems in surface water bodies, and what is the process called?

____

____

____

12. What is channelization?

____

____

____

____

13. What two components of the Earth's atmosphere are the most variable in both time and place?

____

14. What is the difference between weather and climate?

____

____

ANSWERS

1. b
2. c
3. a
4. c
5. zone of aeration; zone of saturation
6. drainage basin
7. percolation
8. F
9. T
10. F
11. When there is excess organic material in a water body, the water becomes clogged with decaying organic material. The decaying material uses up the dissolved oxygen in the water, making it difficult for aquatic life-forms to exist there. This process happens naturally when a lake turns into a swamp; it is called eutrophication.
12. Channelization refers to the modification or engineering of river channels. The modifications usually consist of some combination of straightening, deepening, widening, clearing, or lining of the natural channel for the purposes of flood control, to increase access to floodplain lands, to facilitate transportation, to enhance drainage, or to control erosion.
13. Aerosols and water vapor.
14. Weather is the local condition of the atmosphere (temperature, air pressure, winds, precipitation, etc.) at any given time. Climate is the weather patterns of a particular area averaged over a long period of time.

KEY WORDS

aerated (unsaturated) zone
aerosols
air
aquifer
barrier island
beach drift
channelization
climate
climatologist
condensation
Coriolis effect
deep zone (ocean)
discharge
divide

drainage basin
eutrophication
evaporation
Ferrel cells
flood-frequency curve
glaciation
greenhouse effect
Hadley cells
humidity
hydrologist
infiltration
insolation
interglacial
limnology
longshore current
mesosphere
meteorologist
Milankovitch cycle
monsoon
oceanographer
ozone layer
paleoclimatology
pause (atmospheric)
peak (crest)
percolation
permeability
polar cells
porosity
precipitation
radiatively active (greenhouse) gas
recharge
recurrence interval
reef
salinity
saturated zone
stratosphere
subsidence
surf
surface runoff
surface zone (ocean)
thermocline
thermosphere
tide
troposphere
water table
wave base
wave-cut bench (terrace)
wave-cut cliff
wavelength
weather

10 The Record of Life on Earth

Sufficient for us is the testimony of things produced in the salt waters and now found again in the high mountains, sometimes far from the sea.

—Leonardo da Vinci

These rocks, these bones, these fossil ferns and shells,

Shall yet be touched with beauty, and reveal

The secrets of the book of Earth to man.

—Alfred Noyes

OBJECTIVES

In this chapter you will learn

- what the early Earth was like, and how life may have originated on this planet;
- how the remains of organisms become transformed into fossils;
- how evolution and natural selection work;
- how the history of life is preserved in the fossil record.

1 EARLY EARTH

The history of life is closely intertwined with that of the atmosphere-hydrosphere system. Without a hospitable atmosphere and hydrosphere, life as we know it could not survive. Without life, the atmosphere and the ocean also would not exist in their present forms.

More than 4 billion years ago, in the Hadean eon (see Figure 3.3, page 54), the chemical composition of the atmosphere was very different than it is now. The atmosphere probably consisted of water vapor, carbon dioxide, and nitrogen, with some sulfur compounds and hydrogen chloride. There was no free oxygen (O_2), a necessity for most forms of life on Earth today. It was also very hot, even though the Sun's luminosity (brightness) was lower than it is today. The early atmosphere was composed primarily of greenhouse gases, which trapped heat near the surface. It was too hot for water to exist as a liquid, so there were no oceans, lakes, or rivers. Atmospheric pressure was also much greater than it is today. Altogether, the early Earth was inhospitable to life as we know it.

As the Earth cooled, water vapor began to condense. It fell as rain and collected in low-lying areas, forming bodies of water on the surface, probably as early as 4.4 billion years ago. The early rain was highly acidic because the water reacted with gases in the atmosphere to form acids. The acidic rain reacted with the rocks of the crust, causing chemical weathering. Through reactions with minerals, the acidic rainwater was slowly neutralized. Sediments (the products of chemical weathering of the crust) began to form. Thus, the compositions of the atmosphere, the hydrosphere, and the lithosphere all began to change as materials were exchanged among them.

Why was the rain highly acidic early in Earth history?

Answer: Because the water reacted with gases in the atmosphere to form acids.

2 WHERE DID THE ATMOSPHERE COME FROM?

Earth's original envelope of gases—its **primary atmosphere**—was lost early in Earth history, stripped away by strong solar winds. Little by little, the planet generated a **secondary atmosphere** by releasing volcanic gases from its interior. The main constituent of volcanic gas is water vapor, with varying amounts of nitrogen, carbon dioxide, hydrogen, sulfur dioxide, chlorine, hydrogen sulfide, methane, ammonia, carbon monoxide, and other gases. Overall, the volume of volcanic gases released over the past 4 billion years or so is thought to be large enough and in approximately the right proportions to account for the entire volume of the oceans and atmosphere, with one important exception: the atmosphere is approximately 21 percent oxygen. Where did this abundant oxygen come from?

Some oxygen was probably generated in the early atmosphere by the breakdown of water molecules (H_2O) into hydrogen and oxygen as a result of interactions with ultraviolet radiation. This is an important process, but it doesn't come close to accounting for the present level of oxygen in the atmosphere. Another oxygen-producing process was required. In **photosynthesis,** light energy is used to make carbon dioxide react with water, producing carbohydrates and releasing oxygen. Most organisms in the biosphere depend upon photosynthesis, either directly or indirectly, to obtain food. Almost all of the free oxygen currently in the atmosphere originated through photosynthesis. (You can learn more about photosynthesis and the other biological processes discussed in this chapter if you read *Biology: A Self-Teaching Guide,* by Steven D. Garber.)

Eventually, enough oxygen was created through photosynthesis by early oxygen-producing organisms to permit oxygen to build up in the atmosphere. Along with the buildup of molecular oxygen (O_2) came an increase in ozone (O_3). When ozone began to function as a screen to filter out harmful ultraviolet radiation, organisms were finally able to survive and flourish in shallow waters and, eventually, on land. This critical stage in the evolution of the atmosphere was reached around 600 million years ago. The fossil record shows that there was an explosion of life forms at that time, the transition from Precambrian time to the Phanerozoic eon.

The biosphere also had a profound impact on the carbon content of the atmosphere and ocean. The shells of marine organisms are composed primarily of calcium carbonate ($CaCO_3$), providing a storage reservoir for carbon dioxide (note that $CaCO_3 = CaO + CO_2$). When these organisms die, their shells are buried by seafloor sediments and are eventually transformed into limestone. Limestone is a long-term reservoir for carbon dioxide, isolating it from the atmosphere and hydrosphere. If all the carbon dioxide currently stored in limestone and other sedimentary rocks were released, there would be as much CO_2 in the Earth's atmosphere as in the atmosphere of Venus, where the greenhouse effect runs rampant and the surface temperature is 480°C (900°F).

Where did the free (molecular) oxygen in the atmosphere come from?

Answer: Some of it came from the breakdown of water molecules in the atmosphere as a result of interactions with ultraviolet radiation, but most of it came from photosynthesis.

3 EARLY LIFE

Life played an important role in the chemical transformation of the Earth's atmosphere and hydrosphere. But where and how did life begin? No one knows the complete answer to this question. The first step may have been **chemosynthesis,** the synthesis from inorganic material of small organic molecules such as amino acids, the basic building blocks of proteins. In 1923, a Russian scientist, Aleksandr Oparin, hypothesized that simple organic compounds may have been synthesized from gases in the primitive atmosphere, with energy supplied by lightning or by ultraviolet radiation from the Sun. Thirty years later an American scientist, Stanley Miller, carried out an experiment to test this hypothesis. He passed electric sparks through a mixture of gases similar to those present in the early atmosphere, and recovered some amino acids and other organic compounds. In later experiments all of the important protein-forming amino acids were synthesized, along with other biologically important compounds.

Oparin believed that the organic molecules from which life originated collected as a "soup" in surface waters. Scientists today have many different ideas about where the organic molecules may have come from originally, but most experts agree with at least some aspects of the "soup" hypothesis. But in what environment was this soup formed? Charles Darwin envisioned life as originating in a "warm little pond" on land. Hydrothermal vents at seafloor spreading centers (called "black smokers") or hot springs like those at Yellowstone Park could have provided the raw materials and the heat needed for chemosynthesis. Even the bubbles in sea foam could have provided a site for the collection of early organic material. In fact, the synthesis of organic molecules may have happened in numerous locations, only to be wiped out by meteorite impacts or harsh environmental conditions, before life finally managed to take hold.

Wherever the first organic molecules came from, the big problem is in the next steps: How did the early molecules link together to become the larger "life" molecules? And how did these molecules develop the capability to grow, metabolize, and reproduce? **Biosynthesis** is the polymerization (linking together, as used in mineralogy; see chapter 2) of small organic molecules to form larger organic molecules, including proteins. When amino acids polymerize to create proteins, water is eliminated. This presents a problem for the soup hypothesis, because it would be difficult—if not impossible—to achieve in an aqueous environment. However, if amino acids are dehydrated and heated, polymerization can occur. Perhaps biopolymers were formed when some organic soup along ancient shorelines dried out and was heated by solar radiation or volcanic heat.

Biosynthesis can make larger molecules out of smaller ones; this provides a foundation for growth, one of the fundamental characteristics of living organisms. But biopolymers have no innate mechanism for replicating themselves or for creating new types of molecules. How did these larger organic molecules develop such a mechanism?

You may recall from biology classes that the plan of a living organism is encoded in its **DNA** (deoxyribonucleic acid), a biopolymer that consists of two twisted, chainlike molecules held together by organic molecules. The information and instructions stored in DNA are decoded and executed by **RNA** (ribonucleic acid). Proteins cannot reproduce without RNA because the RNA contains the information required to construct an exact duplicate of the protein molecule. Both DNA and RNA are crucial for the replication of modern cells, but most experts agree that there must have been an early phase in the development of life in which RNA alone was sufficient. The phrase "RNA world" has been coined to describe this phase. Once RNA became established, its presence became the basis for the synthesis and replication of protein molecules.

Another crucial characteristic of life is **metabolism,** the set of chemical reactions through which an organism derives food energy. Organisms that produce their own organic compounds from inorganic chemical compounds are called **autotrophs.** Most autotrophs produce food in the form of carbohydrates, through the process of photosynthesis. The oxygen that was produced by the first photosynthetic autotrophs was toxic to them—a waste product that had to be removed. These organisms required an **anaerobic,** or oxygen-depleted, environment.

Organisms that derive food energy by feeding on other organisms or on organic compounds produced by other organisms are called **heterotrophs.** When heterotrophs consume another organism, the energy stored in the organic compounds is released by one of two processes. Organisms that cannot tolerate oxygen obtain their energy through the anaerobic process of **fermentation,** in which carbohydrate molecules are partially decomposed to form alcohol, carbon dioxide, and water, releasing energy. Heterotrophs that are not anaerobic obtain their energy through the **aerobic** process of **respiration,** which means that they use oxygen to oxidize carbohydrates, creating carbon dioxide, water, and energy.

Respiration is a more efficient process than fermentation, which does not use all available energy. The end product of fermentation, alcohol, is a high-energy compound that can still be used as a fuel; the products of respiration—carbon dioxide and water—cannot. The advent of organisms that tolerated oxygen therefore meant a huge increase in the efficiency with which organisms obtain food energy. This metabolic efficiency allowed for the development and support of more complex structures. It also meant that organisms could grow larger and join in colonies, because they no longer needed a large, free surface area through which to rid themselves of the waste oxygen they were producing.

What is the difference between a heterotroph and an autotroph?

____

____

Answer: Autotrophs produce their own organic compounds from inorganic chemical compounds. Heterotrophs derive food energy by feeding on organic compounds produced by other organisms.

4 CELLS AND CELL PROCESSES

Let's review the essential features that distinguish living from nonliving things. Living organisms can grow, reproduce themselves, and metabolize. Crystals can grow, but they lack the other characteristics of life. Viruses also can grow, but they use the reproductive "machinery" of other organisms in order to replicate themselves, and they lack the ability to metabolize; viruses are "not quite" alive.

All living organisms are composed of one or more cells. The **cell** is the basic structural unit of life, a complex grouping of chemical compounds and structures enclosed in a porous wall, or membrane. The development of the cell membrane was a crucial step in the evolution of life. The membrane separates the materials and chemical reactions that occur inside the cell from the environment outside it. This makes it possible for local organization within the cell to increase. The porous membrane also facilitates the exchange of materials and energy between the cell and its environment. Many bacteria are unicellular (one-celled), but most other organisms are multicellular (more than one cell). Cells may be small (0.01 mm, or 0.0004 in) or large (a few cm, or even larger in rare cases), but whatever their size, all cells are either prokaryotic or eukaryotic.

The earliest cells were **prokaryotic cells** (from the Greek *pro,* "before," and *karyote,* "nucleus," hence "before a nucleus"). Prokaryotes are small, simple cells. The main body of the cell, the cytoplasm, lacks distinctly defined areas in which the cell's various functions are carried out. Most important, the nucleus—the portion of the cell that houses the genetic information—is not separated from the cytoplasm by a membrane. Present-day bacteria and related organisms are prokaryotes. Some prokaryotes are heterotrophs and some are autotrophs, but all autotrophs are prokaryotes. The first oxygen-producing organisms were photosynthetic, thermophilic ("heat-loving"), prokaryotic autotrophs, probably similar to modern cyanobacteria (blue-green algae).

Eukaryotic cells (from the Greek *eu,* "good" or "true," hence, "with a true nucleus") are larger and more complex than prokaryotic cells. Their DNA is housed in a well-defined nucleus that is separated from the cytoplasm by a membrane. The cytoplasm contains a variety of well-defined parts, called organelles, each of which has a specific function. Humans, animals, plants, fungi, and many other living things consist of eukaryotic cells.

All living organisms that are not prokaryotes belong to one of four kingdoms of eukaryotes: Protoctista (single-celled and simple multicellular eukaryotes), Fungi (mushrooms, lichens, and their relatives), Animalia (multicellular organisms that obtain their food by consuming other organisms), and Plantae (multicellular, sexually reproducing eukaryotes that produce their own food; more complicated than algae). The kingdoms are further subdivided, in a hierarchical manner, into successively narrower categories: phylum, class, order, family, genus, and species. A modern human is classified as follows: kingdom: Animalia; phylum: Chordata; class: Mammalia; order: Primates; family: Hominidae; genus: *Homo;* species: *Homo sapiens.*

What is the difference between prokaryotic and eukaryotic cells?

____

____

____

____

Answer: In prokaryotes, the cytoplasm lacks distinctly defined organelles and the nucleus is not separated from the cytoplasm by a membrane. Eukaryotes are larger and more complex; their nucleus is separated from the cytoplasm by a membrane, and their cytoplasm contains well-defined organelles, each of which has a specific function.

5 SPECIES, EVOLUTION, AND NATURAL SELECTION

What were the mechanisms through which simple single-celled and multicellular organisms eventually diversified into the vast array of organisms that inhabit the Earth today? In the ongoing search for an answer to this question, the most significant early contribution was made by Charles Darwin, whose work revolutionized biology and paleontology. On December 27, 1831, Darwin departed from England aboard HMS *Beagle.* When he set sail, Darwin believed in biblical creation and the fixity of species. By the time he returned, his views had changed considerably.

Darwin kept scrupulous notes and made paintings and sketches of the plants, animals, and fossils he saw on the voyage. He was particularly impressed with the many species of finches he saw on the Galápagos Islands. The South American mainland, just a short distance away, had only one species of finch. Darwin reasoned that long ago finches from the nearby mainland had colonized the islands and had subsequently changed as a result of having to adapt to their new environment. In 1859, many years after his return to England, Darwin published his observations and ideas in a book called *On the Origin of Species by Means of Natural Selection*. He waited a long time to publish his findings because he was concerned about the uproar it might—and did—cause. In his book, Darwin outlined the theory of **evolution,** which basically says that new species evolve from old species. (Recall from chapter 1 that a theory is a hypothesis that has been tested and supported by experimentation and observation. Plate tectonics is a unifying theory in geology. Evolution is a unifying theory in biology and paleontology; it draws together a very large amount of information into a simple, testable model.) All present-day organisms are descendants, through a gradual process of adaption to environmental conditions, of different kinds of organisms that existed in the past. Evolution is an essential characteristic of living things, just like growth, metabolism, and reproduction.

Darwin was not the first to suggest evolution as an explanation for the variety and distribution of species. But he was the first to provide a thorough discussion, supported by clear evidence gathered during his voyage and through subsequent research. Moreover, *On the Origin of Species* was published at a time when scientific thought about the Earth was changing rapidly. The concept of uniformitarianism (see chapter 3) was widely accepted; there was a growing consensus that the Earth was very old. And it had been conclusively shown that some plant and animal species that had once lived were now extinct. Most important, Darwin was the first to propose a reasonable mechanism through which evolution could be achieved.

Darwin (and, almost simultaneously, a young naturalist named Alfred Wallace) proposed that evolution could be achieved through the process of **natural selection,** in which poorly adapted individuals tend to be eliminated from a population. When this happens, there are fewer descendants to inherit the genetic characteristics of the poorly adapted individuals and pass them on to the next generation. All natural populations have individuals with varied characteristics. At any one time and in any given environment, some of these characteristics will be more advantageous than others; they will enable the individual to compete more effectively for scarce resources or to escape predators more easily. These individuals are more likely to survive and to have offspring with similar characteristics. This is informally called "survival of the fittest." Over time the entire population evolves as natural selection favors individuals that are particularly well adapted to their environment.

A population's characteristics may diverge when part of the group is subjected to new environmental conditions. This can happen, for example, if part of a population becomes geographically isolated from the rest (Figure 10.1). A rising mountain chain or an invasion by the sea might provide a physical barrier that separates two groups. Or some individuals might migrate across a large river or to an island and thus become isolated from the main population. In such cases the separated group must adjust to a new and different environment. Natural selection will favor those individuals that are best suited to the new environment. Before the separation occurred, the two groups were part of a single **species,** a population of similar individuals that can interbreed and produce fertile offspring. After the separation they may eventually become so different that they can no longer interbreed successfully. In such a case, a new species develops.

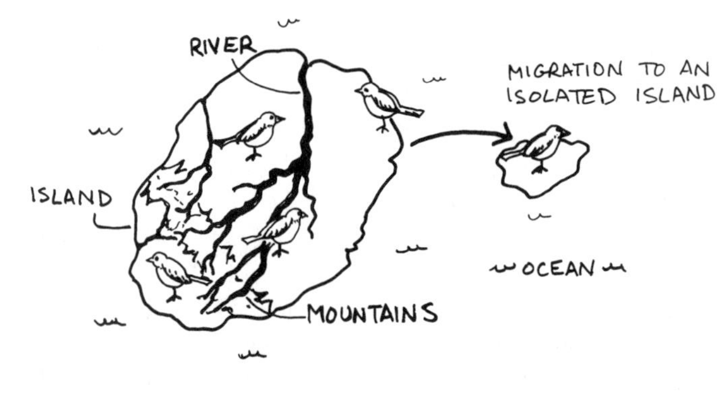

Figure 10.1

What is a species?

Answer: A population of similar individuals that can interbreed and produce fertile offspring.

6 HOW TO BECOME A FOSSIL

In chapter 3, a fossil is defined as the remains of an organism that died and became preserved and incorporated into sediments. Although fossilization can occur in many ways, it is much more common for a newly deceased organism to be destroyed. If the organism is exposed to running water, air, scavengers, or bacteria, it will decompose or be eaten or be broken into parts and scattered around. Hard parts such as bones, teeth, and shells are less easily destroyed and hence more likely to be preserved than soft or delicate parts like skin, hair, leaves, flesh, eggshells, or feathers. In any case, for an organism to be preserved as a fossil it must be quickly covered up by a protective layer of sand or mud, or—in rare cases—tree sap, ice, or tar.

"PETRIFIED WOOD." Fossilized tree trunks in Arizona, once buried, have been exposed by erosion. In the fossilization process, the wood is replaced by silica carried in solution by groundwater, a process called permineralization. (Courtesy U.S. Geological Survey)

Sometimes an organism is preserved with little or no alteration. For example, insects millions of years old have been trapped almost intact in tree sap, which recrystallizes to form the mineral amber (hence the premise for the movie *Jurassic Park*). Organisms can also be preserved if they are trapped in ice or tar, like the ancient wooly mammoths found frozen in Siberia, or the unfortunate prehistoric animals that fell into the La Brea Tar Pits in Los Angeles, California. Animals may also be preserved by natural mummification, in which the soft parts dry and harden before the organism is buried by sediment.

Visit a store that sells jewelry or gems and minerals, and ask if they have any amber. Use a small hand lens to examine the amber. See if you can spot any small fragments of plants or insects preserved in the amber.

More often, the remains of organisms are preserved in an altered state. Bones and other hard parts are sometimes replaced by minerals carried in solution by groundwater. Wood that has been preserved in this manner is called "petrified"

TRACE FOSSILS. Dinosaur footprints near Wethersfield, Connecticut, show where a group of dinosaurs, both adults and juveniles, crossed a muddy bank. Footprints are an example of trace fossils. (Courtesy Peabody Museum, Yale University)

wood. Infiltrating minerals fill tiny pore spaces in the bones, teeth, wood, or shell, strengthening and hardening them. This is called **permineralization.**

If an organism itself is not preserved, it may still leave behind some evidence of its existence. This may be an imprint or mold in the soft sediment that covered it. Molds can reveal fine details long after the organism itself has been destroyed. The delicate remains of plants are sometimes preserved when volatile material in the plant evaporates, leaving an impression in the form of a thin film of carbon. Organisms can leave behind other types of evidence, called **trace fossils.** Footprints are a common form of trace fossil. Worms often leave burrows or borings. Dinosaurs and birds may leave nests containing the remnants of eggshells. Some prehistoric animals even left behind feces that provide clues about their characteristics, habits, and diets.

In what ways can an organism be preserved with little alteration as a fossil?

Answer: By natural mummification (drying) or by being encased in tree sap, ice, or tar.

7 THE MOST ANCIENT FOSSILS

Much of the history of life is the history of unicellular and simple multicellular organisms—that is, microbial life. Indeed, the geologic record of life until the end of Precambrian time—a period of almost 3 billion years—is dominated by **microfossils,** fossils so small that they must be studied under a microscope. The study of such fossils is called **micropaleontology.**

The most ancient fossils that have been found are about 3.55 billion years old. The "chemical signatures" of biological processes have been detected in rocks as old as 3.9 billion years. This suggests that life originated soon after the formation of the Earth, even while the planet was still being bombarded by meteorites. Some of the oldest fossils are the remains of microscopic prokaryotes. Others consist of thin layers of calcium carbonate precipitated from seawater as a result of the action of blue-green photosynthetic bacteria (also prokaryotes). The layered structures, called **stromatolites,** are not the remains of actual organisms, but they provide evidence of the presence of organisms. Similar structures are still being formed today.

For at least 2 billion years, through the Hadean and Archean eons and part of the Proterozoic, the only life on the Earth was prokaryotic. A variety of prokaryotic cells developed (although only one type, cyanobacteria, carried out oxygen-producing photosynthesis). How and where the first eukaryotes originated is a subject of much speculation. We can be reasonably sure that eukaryotes arose from prokaryotes. The chemical pathways in the two classes of cells are so similar that

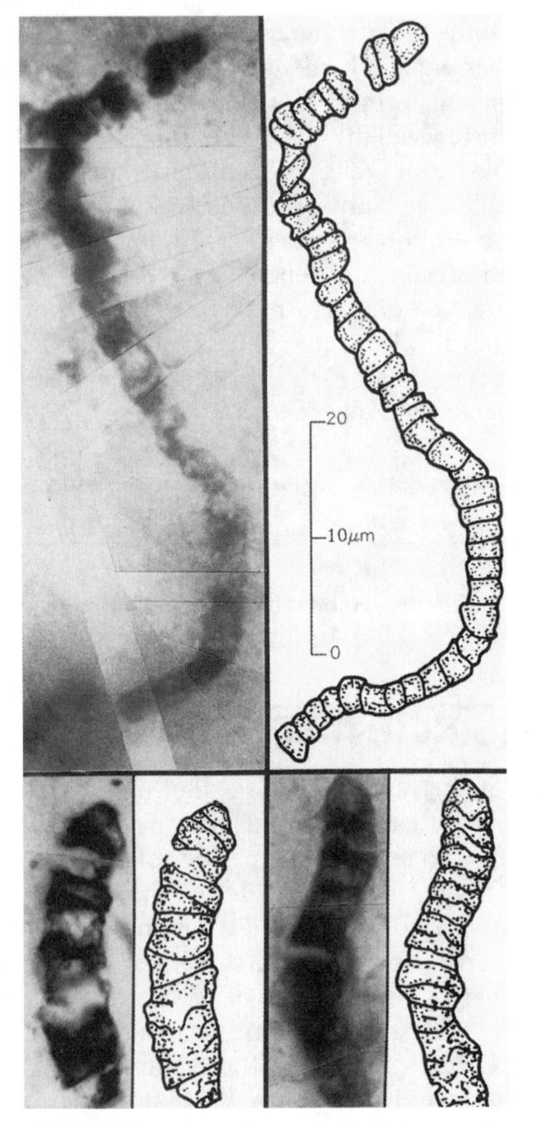

THE MOST ANCIENT FOSSILS. These are examples of the most ancient fossil prokaryotes ever found. They came from a 3.5-billion-year-old rock in Western Australia. Adjacent to each photo is a sketch. The magnification is indicated by the scale: 10 µm (micrometers) = 0.01 mm = 0.0004 in. (Courtesy William Schopf, Department of Earth and Planetary Sciences, UCLA)

they must be related. Most experts believe that eukaryotes originated through a process in which larger prokaryotes enclosed smaller ones.

Once the atmosphere was oxygenated, the emergence of eukaryotes became more likely, for a number of reasons. Prokaryotes need free space around them; crowding interferes with the movement of nutrients and water into and out of the cell. Aerobic eukaryotes are not bothered by crowding, so they can form three-dimensional colonies of cells. Eukaryotes (with the possible exception of the very earliest forms) use oxygen for respiration. As discussed above, this is more efficient than the anaerobic process of fermentation, so eukaryotes do not require as large a surface area as prokaryotes to facilitate the movement of food and waste. This is why eukaryotic cells can be larger (that is, have a greater volume-to-surface ratio) than prokaryotic cells. Because of their greater efficiency in metabolism, eukaryotes can maintain a more complex cell structure.

Eukaryotes appeared at least 1.4 billion years ago, in the middle of the Proterozoic eon. The exact date is not known with certainty, but the fossil evidence clearly shows that by 1 billion years ago they were well established (Figure 10.2). With the appearance of eukaryotes and the transition to an oxygenated atmosphere, more habitats became available, and many new life-forms emerged.

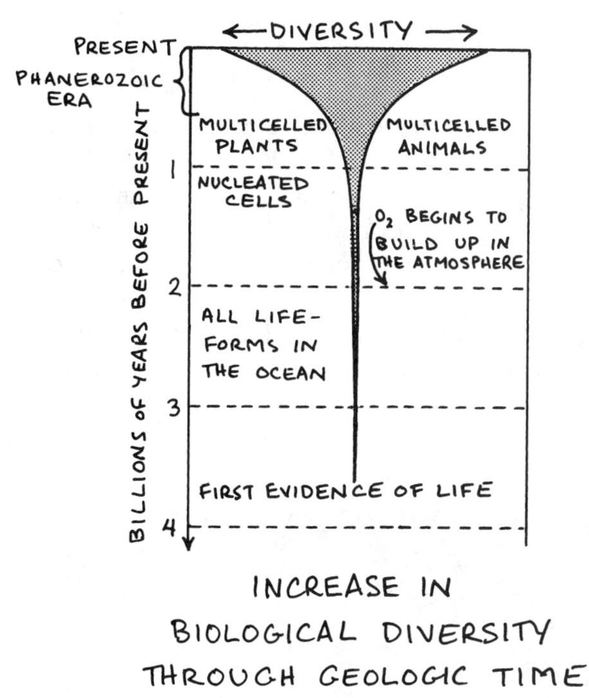

Figure 10.2

The earliest fossils of larger multicellular organisms appear just at the end of the Proterozoic eon in rocks that are about 600 million years old. These fossils, which are now known from a number of localities worldwide, were first found in the Ediacara Hills of South Australia and are called the Ediacara fauna. The Ediacara fauna lived in quiet marine bays. They were jellylike animals with no hard parts. The Ediacara animals represent a huge jump in complexity from the first unicellular eukaryotes, which appeared 800 million years earlier. Scientists still do not know much about what happened during those 800 million years.

What are the oldest known fossils?

Answer: Remains of microscopic prokaryotes (about 3.55 billion years old) and stromatolites.

8 BIOLOGICAL DIVERSITY IN THE CAMBRIAN PERIOD

About 600 million years ago, with the appearance of larger multicellular animals like the Ediacara fauna, life-forms began to diversify very rapidly (Figure 10.2). The Phanerozoic eon, starting with the Cambrian period, was a time of increasing **biodiversity,** that is, increasing number and variety of species. Why was this so? One hypothesis is that sexual reproduction, which developed with the eukaryotes, caused the Cambrian explosion of biological diversity. Another hypothesis is that before that time there was too little oxygen in the atmosphere to support the metabolism of larger organisms. As noted earlier, the increase of ozone (O_3) in the atmosphere shielded Cambrian life-forms from harmful ultraviolet radiation. The rising oxygen content of the atmosphere may also have affected the biochemistry of calcium phosphate and calcium carbonate, the two main skeleton- and shell-building components.

Whatever the reasons, a great many changes occurred about 550 million years ago in the early Cambrian. Compact animals were evolving to replace the soft-bodied, jellylike organisms of Ediacara times: trilobites, mollusks (clams and sea snails), and echinoderms (sea urchins). All of these (except trilobites) are types that have persisted up to the present. These animals were equipped with gills, filters, efficient guts, a circulatory system, and other characteristics of more advanced life-forms. The Cambrian also saw the development of skeletons, internal and external. These gave organisms a selective advantage, protecting them against predators, against drying out, against being injured in turbulent water, and so on. Many soft-bodied creatures also persisted in the Cambrian. They can be seen in the fossils of the Burgess Shale, a beautiful collection of soft-bodied animals and plants that were covered by black muds in Cambrian times and eventually preserved as fossils in

ANCIENT FOSSIL. *Dickinsonia costata,* the remains of a creature that lived about 600 million years ago. Like a jellyfish, *Dickinsonia* lacked hard parts, so the fossil is an impression made on sand when the creature died. The specimen is 3 inches in length and came from the Flinders Range, South Australia. Photograph by William Sacco.

shale in British Columbia. The rich diversity and extraordinary life-forms of the Cambrian marine environment are almost unparalleled in the history of the Earth.

Why do you think early geologists chose the beginning of the Cambrian period as the dividing line between Proterozoic time and Phanerozoic time?

____

____

Answer: There was an enormous explosion of biological diversity at the beginning of the Cambrian; this is preserved in the fossil record as a distinctive transition.

The great proliferation of life in the Cambrian was confined to the sea. Successful organisms diversified and flourished; unsuccessful ones disappeared. By 500 million years ago the main kinds of structural organization for animal life had been established. The big step that remained was to leave the sea and occupy the land. Eventually all kingdoms of life took that step.

The requirements for life on land are the same for all organisms. The most important are:

1. Structural support, needed because aquatic organisms are buoyed up by water, whereas land organisms must contend with gravity
2. An internal aquatic environment, with a plumbing system and methods of conserving water against loss to the surrounding atmosphere
3. A method of exchanging gases with air instead of with water
4. A moist environment for the reproductive system, essential for all sexually reproducing organisms

9 PLANTS

The earliest land plants evolved from green algae. Eventually, vascular plants evolved, with structural support from stems and limbs (requirement 1), and a set of channels through which water and dissolved elements are transferred from roots to leaves (requirement 2). Gas exchange with the air (requirement 3) is controlled by adjustable openings (stomata) in the leaves.

The earliest land plants were seedless. Ferns and club mosses are modern seedless plants. Many of the adult forms of seedless plants can tolerate some drought, but all plants rely on moisture for the sexual phase of their reproductive cycles. Without moisture, the reproductive cells have no medium in which to reach each other and fuse, so fertilization does not occur. Consequently, seedless plants have never been able to survive in places that lack a dependable supply of moisture for at least part of the growing season. Seedless plants reached their peak in the Mississippian and Pennsylvanian periods, when they dominated the vast forests on the tropical floodplains and deltas of North America, Europe, and Asia. The remains of these plants produced huge deposits of coal.

By the middle Devonian a few plants were on the way to meeting requirement 4, providing their own moist environment to facilitate sexual reproduction. These plants were the **gymnosperms** ("naked-seed plants"). The female cell of a gymnosperm is attached to the vascular system and therefore has a supply of moisture. The male cell is carried in a pollen grain with a waxy coating. When the two fuse, a seed results. The seed provides moisture and nutrients that sustain the growth of the young plant until it can support itself through photosynthesis. This important change allowed plants to survive in other habitats besides swampy lowlands. Naked-seed plants survive today; ginkgoes and conifers are examples. Freed from their original swampy habitat, the gymnosperms did not have to compete with the great seedless trees of the coal forests. By the end of the Pennsylvanian period, they had spread over most of the world.

Gymnosperms have one drawback. The male cell carrier, the pollen, is spread through the air. What chance does a pollen grain in the air have of finding a female cell? The odds are against it. To ensure success, gymnosperms have to make huge amounts of pollen. Flowering plants (**angiosperms,** or enclosed-seed plants) solved this problem. For a small incentive (nectar or a share of the pollen), insects deliver the pollen from one flower to another, or from one part of the plant to another. Birds and other animals also help the angiosperms by eating their seed-bearing fruits. The seeds are distributed throughout the animal's territory in its feces. Angiosperms evolved after gymnosperms, but by the end of the Cretaceous period angiosperms had become the dominant land plants. Angiosperms have developed close relationships with animals: insects for pollination, and birds and quadrupeds (four-footed animals) for seed dispersal. Many flowering plants, such as grasses and birches, also rely on the wind to help disperse their seeds and pollen.

What is the difference between angiosperms and gymnosperms? Which developed first?

___

___

Answer: Angiosperms are flowering, enclosed-seed plants. Gymnosperms are naked-seed plants; they developed first.

10 ANIMALS

The first major expansion of multicellular marine animals occurred in the last few million years of Precambrian time, at the end of the Proterozoic eon. Among the creatures in the Cambrian seas were many that belong to the phylum Arthropoda, so called because of their jointed legs. Modern arthropods include crabs, spiders, centipedes, and insects, members of the most diverse phylum on the Earth. Arthropods were the first animals to make the change from sea to land.

With a few exceptions, arthropods were quite small and light. They were covered with a hard shell of chitin (a fingernail-like material). Thus, they were well adapted for life on land in regard to structural support and water conservation. The first to go on land were probably Silurian centipedes and millipedes. Insects were abundant by the Mississippian period, and included dragonflies with a wingspan of up to 60 cm (24 in). For all their success as land creatures, however, the arthropods have very primitive respiratory and vascular systems. For example, insects breathe through tiny tubes that penetrate the outer coating. This mode of respiration severely limits the size of an organism and is the reason why most insects are small. Their "blood" is simply body fluid bathing the internal organs; it does not circulate in closed vessels. The fluid is kept in motion by a sluggish "heart" that is little more than a contracting tube. At first it seems odd that these primitive animals could have diversified into more than a million terrestrial species. Yet the arthropods' simple vascular system is obviously effective.

Among the fossils of the Burgess Shale is a small, inconspicuous fossil called *Pikaia*. *Pikaia* is a **chordate** because it has a notochord, a cartilaginous rod running along the back of the body. (Humans are also chordates. We have a notochord as embryos; later it is replaced by the backbone.) *Pikaia* and other Cambrian fish were jawless, probably feeding on organic matter dredged from the seafloor. In these jawless fish we see the first important stage in the development of **vertebrates,** animals that possess backbones. Jawed fish came next. With the development of jaws came a great burst of diversification. The original jawless fish, only a few centimeters long, were quickly joined by larger fish. These included 9-meter (29.5-ft) armored sharks and other cartilaginous fish, and the huge order of ray-finned fish that are familiar to us as game and food fish. The possession of jaws allowed fish to move into a wide range of ecological niches.

The first fish to venture onto land, in the Devonian period, may have been one of the Crossopterygii, or lobefinned fish. (Interestingly, crossopterygians were thought to be extinct until a living example called *Coelacanthus* was found in the Indian Ocean in 1938.) Crossopterygians had several features that could have enabled them to make the transition to land. Their lobelike fins, for example, contained all the elements of a quadruped limb. They had internal nostrils, a characteristic of air-breathing animals. As fish, they already had developed a vascular system that was adequate for life on land. However, recent DNA studies of modern coelacanths, amphibians, and lungfish suggest that lungfish, rather than lobefinned fish, may have been the ancestors of the first land-dwelling amphibians. The lungs used by modern lungfish to take occasional gulps of air during periods of drought were developed long before the first amphibians appeared. Presumably, it would have been a relatively simple adaptation for these lungs to become full-time suppliers of air for amphibians in a terrestrial environment.

Amphibians never developed an effective method for conserving water. To this day they have permeable skins, which is one reason they have never become wholly independent of aquatic environments. Although amphibians have been successful for many millions of years, they have one difficulty that limits their ability to expand: they have never met the reproductive requirement for life on land. In most amphibian species, the female lays her eggs in water, the male fertilizes them there after a courtship ritual, and the young are fishlike when first hatched (like tadpoles). Amphibians, with one foot on the land, have remained tied to the water for breeding. Although some became quite large (2 to 3 m, about 3 yd), they never diversified much after the Devonian period. One branch went on to become reptiles. Of the rest, those that survive are frogs, toads, newts, salamanders, and limbless water "snakes," which have returned to life in the water.

Reptiles freed themselves from the water by evolving an egg that could be incubated outside of the adult, and by developing a watertight skin. These two modifications enabled them to occupy terrestrial niches that the amphibians had missed because of their need to live near water. Originating in the Mississippian and Pennsylvanian coal swamps, by the Jurassic period the reptiles had moved over the land, up into the air, and back to the water. They had also produced two orders of dinosaurs (the largest quadrupeds ever to walk the Earth) and given rise to two new vertebrate classes, mammals and birds.

Vertebrate animals made the transition into the air in the form of pterosaurs, flying reptiles with long wings and tails. Birds first appeared near the end of the Jurassic period. An early example of a bird, *Archaeopteryx* ("ancient wing"), would have been classified as a dinosaur were it not for the discovery that it had feathers.

In many ways, mammals are better equipped to live on the land than were the great reptiles. Mammals, mostly quadrupeds, are adapted to a faster and more versatile life than the reptiles. By comparing brain-to-body-weight ratios in archaic and modern reptiles and mammals, it can be shown that increase in mammalian brain size is a continuing process, whereas in reptiles brain size has not increased; the ratios in modern reptiles do not differ significantly from those in archaic ones. Still, it took the extinction of the dinosaurs to allow for the great expansion of mammals at the beginning of the Cenozoic era.

What was the first type of animal to make the transition from sea to land?

Answer: Arthropods.

EARLY BIRD. *Archaeopteryx macroura* is an ancient bird preserved as a fossil in the Solenhofen Limestone of Bavaria, Germany. *Archaeopteryx* lived about 120 million years ago, and was one of the first animals to develop feathers. (Smithsonian Institution)

11 THE HUMAN FAMILY

The family of humans, Hominidae, did not descend from the modern ape family, Pongidae. Both families probably diverged from an earlier apelike family. The fossil record for both humans (**hominids**) and apes is poor; many transitional forms are missing, but new and important finds are made by paleontologists every year. These finds are helping scientists to fill in the history of our human family.

At this writing, the oldest known definitive example of a hominid fossil, dated at 4.4 million years, is *Ardipithecus ramidus.* The genus *Ardipithecus* was succeeded by *Australopithecus,* with six known species (so far), the earliest of which (*Australopithecus anamensis*) dates from about 4.2 million years ago. The australopithecines are probably best represented by *Australopithecus afarensis,* of which the famous "Lucy" fossil is an example. The australopithecines were generally small, standing only about 1.0 to 1.5 m (3.5 to 5.0 ft) in height, but they were physically strong and their brain capacity was larger than that of chimpanzees. From the shape of the pelvis and from footprints left in soft volcanic mud more than 3.0 million years ago, we know that these individuals walked upright, though their skulls still looked more apelike than human. Successive australopithecine species inhabited Africa, each overlapping in time with the preceding species, as shown in Table 10.1; they disappeared altogether about 1.1 million years ago.

Homo erectus was probably the first species of our own genus (*Homo*). Fossils of *H. erectus* dating back about 1.8 million years have been found in Africa, Europe, China, and Java. An even earlier species called *Homo habilis,* "handy man," is thought to have used stone tools. Since toolmaking is the distinguishing feature of the genus *Homo,* some experts include *H. habilis* in this genus; others argue that the skull of *H. habilis* is more like that of the australopithecines.

Homo erectus disappeared around 300,000 years ago. By about 230,000 years ago, *Homo neanderthalensis*—"Neanderthal man"—had appeared. The fossil record between 400,000 and 100,000 years ago is poor, so the transition from *H. erectus* to *H. neanderthalensis* is not well understood. From the study of burial sites as much as 100,000 years old, paleontologists and anthropologists have deduced that the Neanderthal people practiced some form of religion. On the basis of similar-

Table 10.1 Approximate Time Line of Hominid Species

	Millions of Years Ago (MYA)					
Species	**5.0**	**4.0**	**3.0**	**2.0**	**1.0**	**0.0**
Ardipithecus ramidus (4.4 mya)						
Australopithecus anamensis (4.2–3.9 mya)						
Australopithecus afarensis (3.9–3.0 mya)						
Australopithecus africanus (3.0–2.0 mya)						
Austraolopithecus aethiopicus (2.6–2.3 mya)						
Australopithecus robustus (2.0–1.5 mya)						
Australopithecus boisei (2.1–1.1 mya)						
Homo habilis (2.4–1.5 mya)						
Homo erectus (1.8 mya–300,000 ya)						
Homo neanderthalensis (230,000–30,000 ya)						
Homo sapiens (120,000 ya–present)						

ities in teeth and brain size (slightly larger than our own), some experts argue that Neanderthal was part of the modern human species; thus, they label it *Homo sapiens neanderthalensis.* However, recent studies suggest that the DNA of Neanderthal is different from our own, which suggests that *Homo sapiens* is not a direct descendant of the Neanderthal. Neanderthals disappeared about 30,000 years ago; they were replaced by the biologically modern Cro-Magnon people, the first indisputable example of our own species, *H. sapiens.*

Did the Cro-Magnon evolve from the Neanderthal, or were they a distinct species? Both peoples lived in Europe for a period of up to 10,000 years before the disappearance of the Neanderthal. Did they meet? Did they interbreed? Were the Neanderthal wiped out by the Cro-Magnon? These and many other questions await answers, as paleontologists continue to search for clues in the fossil record.

Has there been a time in Earth history when more than one hominid species was alive?

Answer: Yes.

12 MASS EXTINCTIONS

Embedded in the fossil record is a story of adaptation and recovery following catastrophic episodes in which many species become extinct within a geologically short time. Such episodes are called **mass extinctions.** Most people are aware that the dinosaurs became extinct about 65 million years ago, at the boundary between the Cretaceous (K) and Tertiary (T) periods. But many are not aware that other animal and plant species were also affected. Approximately one-quarter of all known animal families living at the time, including marine and land-dwelling species, became extinct at the end of the Cretaceous period. This mass disappearance of species is clearly evident in the fossil record. It is the reason that early paleontologists selected this particular stratigraphic horizon to represent a major boundary in the geologic timescale.

The great K-T extinction is not unique, nor was it the most dramatic of such occurrences. There have been at least 5 and possibly as many as 12 mass extinctions during the past 250 million years. The most devastating of these occurred 245 million years ago at the end of the Permian period, when as many as 96 percent of all species died out. Another great extinction occurred at the end of the Triassic period, and several earlier extinctions affected marine organisms.

What causes mass extinctions? Some evidence suggests that the K-T extinction may have been caused by a giant meteorite impact. If an extraterrestrial body

METEORITE IMPACT CRATER. Meteor Crater, Arizona, has a raised rim, formed as a result of broken rock being thrown out by a meteorite impact. The crater is 50,000 years old, 1.2 km (0.75 mi) in diameter, and 200 m (656 ft) deep. Many much larger impacts have occurred during the Earth's long history. Note that this particular impact was too small and happened much too recently to have caused the extinction of the dinosaurs. (Courtesy U.S. Geological Survey)

such as a meteorite or a comet 10 km (more than 6 mi) in diameter struck the Earth, it would cause massive environmental devastation. The effects could include earthquakes, tsunamis, widespread fires, acid rain, atmospheric particulates that might cause global darkness, and intense climatic changes. Evidence for these and related effects has been found in the stratigraphic horizon that marks the K-T boundary. Throughout the world the boundary is also marked by a thin layer of clay that is rich in the element iridium (Ir). This is consistent with an influx of extraterrestrial material, because meteorites contain a great deal of iridium compared to the amount contained in terrestrial rocks.

It is possible that a meteorite impact caused the K-T extinction, but the causes of other major extinctions are not as clear. Many scientists feel that some extinctions—particularly the great marine extinctions of the Paleozoic era—were more likely caused by climatic or other environmental changes than by catastrophic events such as meteorite impacts. The study of mass extinctions seems particularly relevant today; the present rate of species extinctions from human causes is rivaled only by the greatest mass extinctions in Earth history.

What was the greatest mass extinction in geologic history?

___

Answer: The Permian extinction, 245 million years ago.

SELF-TEST

These questions are designed to help you assess how well you have learned the concepts presented in chapter 10. The answers are given at the end.

1. The most ancient fossils that have been found are about ___ billion years old.
   a. 1.5
   b. 2.5
   c. 3.5
   d. 4.5
2. "Petrified wood" is wood that has been preserved by the process of ___.
   a. natural mummification
   b. carbonization
   c. molding
   d. permineralization
3. Although *Archaeopteryx* had feathers, many scientists feel that it should be classified as a(n) ___ rather than as a bird.
   a. dinosaur
   b. mammal
   c. arthropod
   d. amphibian
4. ___ is the idea that poorly adapted individuals tend to be eliminated from a population, with the result that fewer descendants will inherit the genetic characteristics of the poorly adapted individuals.
5. ___ is the study of fossils that are so small they must be studied with a microscope.
6. The mass extinction that was responsible for the extinction of the dinosaurs (as well as other species) occurred ___ million years ago, at the boundary between the ___ and ___ periods.
7. The early Earth was very hot, because the Sun's luminosity was much brighter than it is today. (T or F)
8. Respiration is more efficient than fermentation as a mechanism for organisms to obtain energy. (T or F)

9. The earliest land plants were the gymnosperms; the angiosperms followed closely thereafter. (T or F)
10. What is the role of limestones and shelled organisms in regulating climate and modifying the chemistry of the Earth's atmosphere?

11. What are trace fossils, and what are some types of trace fossils?

12. Did the family of humans (Hominidae) evolve directly from modern apes (Pongidae)?

13. What is a chordate? Are humans chordates?

ANSWERS

1. c
2. d
3. a
4. Natural selection
5. Micropaleontology
6. 65; Cretaceous; Tertiary
7. F
8. T
9. F
10. Limestone is made from the remains of shelled organisms. Limestone ($CaCO_3$) is a huge, long-term storage reservoir for carbon dioxide, removing it from the atmosphere and hydrosphere. If all the carbon dioxide currently stored in limestones were released back into the atmosphere, the Earth would have a runaway greenhouse effect and an atmosphere like that of Venus, where the surface temperature is about 480°C.

11. A trace fossil is fossilized evidence left behind by an organism, without necessarily leaving any part of the organism itself. Examples include footprints, burrows, borings, nests, molds or imprints, and feces.
12. No, the two groups probably diverged from an earlier apelike family.
13. A chordate is an animal with a notochord, a cartilaginous rod running along the back of the body. Humans are chordates; we have a notochord as embryos, later replaced by the backbone.

KEY WORDS

aerobic
anaerobic
angiosperm
autotroph
biodiversity
biosynthesis
cell
chemosynthesis
chordate
DNA (deoxyribonucleic acid)
eukaryotic cell
evolution
fermentation
gymnosperm
heterotroph
hominid
mass extinction
metabolism
microfossil
micropaleontology
natural selection
permineralization
photosynthesis
primary atmosphere
prokaryotic cell
respiration
RNA (ribonucleic acid)
secondary atmosphere
species
stromatolite
trace fossil
vertebrate

11 Resources from the Earth

They wonder much to hear that gold, which in itself is so useless a thing, should be everywhere so much esteemed.

—Sir Thomas More

Yet it isn't the gold I'm wanting so much as just finding the gold.

—Robert Service

OBJECTIVES

In this chapter you will learn

- how dependent modern society has become on mineral resources;
- how mineral deposits are formed;
- how fossil fuel deposits are formed;
- why it is important to seek alternatives to fossil fuel use.

1 RESOURCES AND MODERN SOCIETY

Minerals and many types of energy resources are **nonrenewable resources**—they cannot be renewed or replenished. The geologic processes responsible for the formation of most types of mineral and energy resources are still operating today, but they may take hundreds of millions of years to be completed. As a society, we can't wait that long for a resource that has been used up to be replenished. Therefore, it is more accurate to say that nonrenewable resources are not renewable or replenishable *on any humanly accessible time scale.* **Renewable resources** such as fish or trees require a different management style from nonrenewable resources. Renewable resources can be replenished, so stocks can be maintained if resources are managed carefully and sustainably. Nonrenewable resources are depleted as fast as we use them, so we must use them sparingly and recycle them whenever possible.

Look around you. Even if you are reading this book under a tree in a meadow, you are surrounded by products made from mineral resources. Your pen is metal or plastic (a petroleum product). Your watch has metal parts, and the face is glass (a silica product) or plastic. Your jewelry, your glasses, even your clothes (if you are wearing synthetic fabrics, which are petroleum products)—all are made from mineral resources. Even the paper in this book has mineral additives to give it texture and color. And virtually every part of the car, bicycle, or public transport vehicle that got you where you are was made from mineral resources found by a geologist and dug out of the ground by miners.

Our hunter-gatherer ancestors once lived entirely on renewable resources—the animals they caught and the plants they gathered. Eventually, several million years ago, our ancestors crossed one of the great thresholds of human development. They picked up stones to use as hunting tools and became the first and only species to use nonrenewable resources routinely. It was not long before they discovered that some stones are tougher and make better spearpoints than other stones. Because the best stones could only be found in a few places, trading started. As time passed, people started gathering and trading salt. Hunting communities can satisfy their dietary needs for salt by eating meat, but when farming started diets became grain based, and extra salt was needed. We don't know when or where salt mining started, but long before recorded history, trade routes for the exchange of nonrenewable resources—beginning with spearheads and salt—crisscrossed the globe.

Metals were first used about 17,000 years ago. Copper and gold are both found in nature as pure **native metals,** and these were the first to be mined for human use. Native copper is rare, and eventually other sources of copper were needed. About 6,000 years ago our ancestors learned how to extract copper from certain rocks by smelting—heating the rocks to separate the molten metal from unwanted waste materials. By 5,000 years ago, they had learned how to smelt lead, tin, zinc, silver, and other metals, and how to mix metals to make alloys such as bronze (copper + tin) and pewter (tin + lead + copper). The smelting of iron is more difficult than the smelting of copper and was only achieved about 3,300 years ago.

The first fuels to be used as energy sources were renewable—wood and animal dung. The Babylonians were the first to use oil, about 4,500 years ago. The oil came from natural seeps in the valleys of the Tigris and Euphrates Rivers, in what is now Iraq. The first people to mine coal and drill for natural gas were the Chinese, about 3,100 years ago. By the time the Roman Empire came into existence in Europe about 2,500 years ago, humans had come to depend on a very wide range of natural resources—not just fuels and metals, but processed materials such as cement, plaster, glass, porcelain, and pottery.

The list of materials we mine, process, and use has grown steadily since then. Today more than 200 different kinds of minerals are mined and used. Each of us uses—directly and indirectly—a very large amount of material derived from nonrenewable resources. Without them we could not build planes, cars, televisions, or computers. We could not distribute electricity or build tractors to till fields and produce food. Machines are used to make clothes, transport us, and help us to communicate. The metals needed to build machines, as well as the fuels needed to run them, are nonrenewable resources extracted from the Earth. We depend fundamentally on nonrenewable mineral and energy resources; without them, modern society would collapse.

When did our ancestors learn about smelting?

Answer: About 6,000 years ago.

2 MINERAL RESOURCES

The diversity of minerals and rocks that provide materials used by humans is so great that to make a simple classification of mineral resources is almost impossible. Nearly every kind of rock and mineral can be used for something, although those that are most valuable tend to be rare. It is convenient to group mineral resources according to how they are used, as in Table 11.1. **Metallic mineral resources** are mined specifically for the metals that can be extracted by smelting. Examples are sphalerite (ZnS), from which zinc is recovered, and galena (PbS), from which lead is recovered. **Nonmetallic mineral resources** are mined for their properties as rocks or minerals, not for the metals they may contain. Examples are halite (salt), gypsum, and clay.

The branch of geology that is concerned with mineral resources and discovering new supplies of useful minerals is **exploration geology** (also called **economic geology**). A mineral **deposit** is a local concentration or enrichment of a given mineral. Exploration geologists seek deposits from which the desired minerals can be recovered least expensively. We use the word **ore** to describe a deposit from which one or more minerals can be extracted profitably.

Table 11.1 Some Examples of Mineral Resources and Their Uses

Metals	
Abundant metals	Iron, aluminum, magnesium, manganese, titanium, and others
Scarce and rare metals	Copper, lead, zinc, nickel, chromium, gold, silver, tin, tungsten, mercury, molybdenum, uranium, platinum, and many others
Nonmetals	
Used for chemicals	Sodium chloride (halite), sodium carbonate, borax, calcium fluoride (fluorite)
Used for fertilizers	Calcium phosphate (apatite), potassium chloride, sulfur, calcium carbonate (limestone), sodium nitrate
Used for building	Gypsum (for plaster), limestone (for cement), clay (for brick and tile), asbestos, sand, gravel, crushed rock, shale (for brickmaking and for cement)
Used for jewelry	Diamond, corundum (ruby and sapphire), garnet, amethyst (quartz), beryl (emerald), and many others
Used for glass and ceramics	Clay, feldspar, silica (quartz sand)
Used for abrasives	Diamond, garnet, corundum, pumice, quartz

All ores are mineral deposits, but not all mineral deposits are ores. Whether or not a given mineral deposit is an ore is determined by how much it costs to extract the mineral, and how much people are prepared to pay for it. Many factors determine whether or not a mineral deposit is an ore. The concentration, or **grade,** of the ore is one important factor. In general, the more highly concentrated the ore is, the higher the grade and the more valuable the deposit. The size of the deposit, its depth, the distance to a road or a railroad, the cost of labor, and the demand for the material are other factors. The nonvaluable minerals associated with ores are called **gangue** (pronounced "gang").

Economically exploitable mineral deposits are distinctly localized within the Earth's crust. To a certain extent, the distribution of different mineral resources can be linked to geologic processes that happen in specific tectonic environments. This permits exploration geologists to identify locations that will likely contain specific types of mineral deposits. Over the past few decades, there has been a slow but steady shift in the emphasis of mineral exploration and production away from the industrialized nations and toward the developing nations. In the industrialized world, the geologic locations that are most favorable for conventional mineral exploration have already been prospected, assessed, and in some cases mined and depleted. This doesn't mean that there are no more mineral deposits to be found; it just means that geologists will have to search harder, utilize innovative exploration techniques, and learn to look for mineral deposits in unconventional locations.

What is the difference between a mineral deposit and an ore?

Answer: A mineral deposit is a local concentration of one or more minerals. An ore is a mineral deposit that contains valuable minerals—that is, minerals that can be extracted profitably.

3 HOW MINERAL DEPOSITS ARE FORMED

For a mineral deposit to form, a geologic process or combination of processes must produce a localized enrichment of one or more minerals. Minerals can become concentrated as a result of the following processes:

1. Hot, aqueous (water-rich) solutions flowing through fractures and pore spaces in rocks
2. Metamorphic processes
3. Magmatic processes
4. Evaporation or precipitation from lake water or seawater
5. The action of waves or currents in flowing surface water
6. Weathering processes

Many famous mines contain ores that formed when minerals were deposited from **hydrothermal solutions**—hot, aqueous fluids. (*Hydro* is the Greek word for "water," and *thermal* is from *therme,* "heat"). Hydrothermal solutions deposit their dissolved constituents in cracks, forming veinlike concentrations of ore minerals. Hydrothermal solutions can form in many ways, but most commonly they are fluids that are released from cooling magma bodies (Figure 11.1). Hydrothermal mineral deposits are the primary sources of many metals, particularly copper, lead, zinc, mercury, tin, molybdenum, tungsten, gold, and silver.

Hydrothermal activity is usually associated with contact metamorphism, the alteration of rocks adjacent to bodies of hot rock or magma, and with metasomatism, the movement of chemical constituents by solutions during contact metamorphism (see chapter 8). Regional metamorphism—the alteration of large areas of rock as a result of high temperatures and pressures associated with tectonism—can also act to concentrate minerals. In response to the differential stresses that occur during regional metamorphism, minerals may become segregated and concentrated into distinct bands or layers. A wide variety of valuable nonmetallic mineral resources are concentrated in banded metamorphic rocks, including micas, asbestos, graphite, and some gemstones. Another common metamorphic

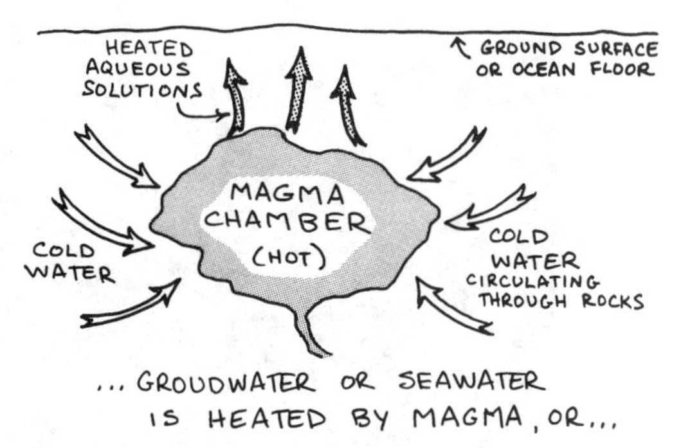

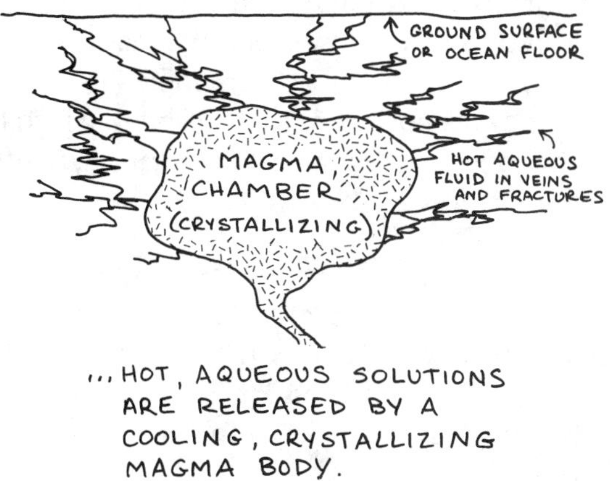

Figure 11.1

process is recrystallization. The pure, white marbles of Carrara, Italy, highly valued by sculptors, began as porous aggregates of shells and other fragments. Through metamorphic recrystallization, the original shells were obliterated and the rocks were transformed into crystalline marbles composed entirely of interlocking grains of calcite.

There are several ways by which ore minerals may become concentrated as a result of magmatic (igneous) processes. For example, when a large chamber of basaltic magma crystallizes, one of the first minerals to form is chromite. Chromite is the main ore mineral of chromium, a constituent of steel. The chromite crystals, which are denser than the liquid around them, settle and then accumulate at the bottom of the magma chamber. This process can produce almost pure layers of chromite. Another important kind of magmatic mineral deposit is kimberlite (see Figure 4.1, page 69), a long, thin, pipelike body of igneous rock made from

CHROMIUM ORE. Chromite, the main ore mineral of chromium, crystallizes from a magma. Because it is denser than the magma, the chromite sinks to the bottom and accumulates in layers. This photograph shows layers of chromite (black) alternating with layers of plagioclase feldspar (white), which formed during the crystallization of the Bushveld Igneous Complex, Dwars River, South Africa. (Courtesy Brian Skinner)

magma that originates deep in the mantle—150 km (about 90 mi) or more. Kimberlite magma rises explosively, transporting broken fragments of mantle rock (xenoliths) as it goes. One of the mineral constituents of the xenoliths in kimberlite is diamond, a high-pressure mineral that forms only at depths greater than 150 km. The only way natural diamonds can reach the surface of the Earth is through kimberlite pipes.

Mineral deposits can form when dissolved substances precipitate from lake water or seawater. One cause of precipitation is evaporation, in which layers of salts are left behind in evaporite deposits. Sodium carbonate, sodium sulfate, and borax come from deposits formed by the evaporation of lake water. Marine evaporites, from the evaporation of seawater, produce gypsum, halite, and a variety of potassium salts used in fertilizers. Another cause of precipitation is biochemical reactions in seawater. Such reactions lead to the precipitation of the mineral apatite, the main source of phosphate fertilizers. In sedimentary rocks older than about 2 billion years, there are unusual iron-rich sedimentary rocks called banded iron formations, which are thought to be ancient biochemical precipitates from seawater. In banded iron formations (discussed in chapter 7), bands of red (oxidized) iron-bearing minerals alternate with bands of black (reduced, or oxygen-depleted) iron-bearing minerals. The alternating bands are thought to reflect the precipitation of iron during a period when oxygen levels in the atmosphere (and thus the ocean, which is in chemical equilibrium with the atmosphere) were fluctuating. (See chapter 10 to review the chemical evolution of the atmosphere and the ocean.)

Heavy mineral grains can be concentrated as a result of the sifting or winnowing action of flowing water. As the water sifts the sediments, the heaviest minerals sink and become concentrated in layers or pods (Figure 11.2). A deposit that forms in this way is called a **placer deposit.** The flowing water may be a stream or a longshore current in the sea. Heavy minerals that typically become concentrated in placer deposits are gold, platinum, diamonds, chromite, and minerals that contain zirconium and titanium.

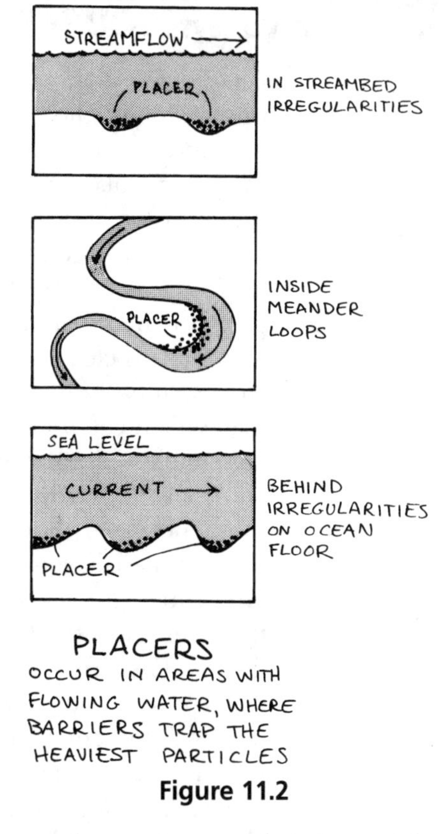

Figure 11.2

Chemical weathering in tropical climates can lead to the concentration of minerals through the removal of soluble materials. The soluble materials are carried downward by infiltrating rainwater, leaving behind a residual concentration of less soluble material. A common example of a **residual deposit** that is formed in this manner is laterite, a hard, highly weathered subsoil rich in insoluble minerals. Laterite is mined for iron and sometimes nickel. Iron-rich laterite is by far the most common kind of residual deposit, but the most important for human exploitation is bauxite, the source of aluminum ore. Bauxite deposits are concentrated in the tropics. Where bauxite is found in present-day temperate conditions, such as France, China, Hungary, and Arkansas, we can infer that the climate was tropical when the bauxite formed.

What is the possible connection between banded iron formations and the chemical evolution of the atmosphere and ocean?

___

___

___

Answer: Alternating bands of red (oxidized) and black (reduced, or oxygen-depleted) iron-bearing minerals in banded iron formations are thought to reflect the precipitation of iron during a period when oxygen levels in the atmosphere and the ocean were fluctuating.

4 ENERGY RESOURCES AND THE ROLE OF FOSSIL FUELS

The Earth's energy comes from three sources: solar radiation, geothermal energy from natural radioactivity within the Earth, and tidal energy from the Earth's rotation and gravitational interactions with the Moon and the Sun. Energy from these sources circulates through the many pathways and reservoirs of the Earth system, driving processes such as photosynthesis and atmospheric circulation. The actual flow of energy across the Earth's surface, all of which could theoretically be used, is $174,000 \times 10^{12}$ watts (W) (174,000 terawatts). Each of the world's 6 billion people uses energy at an average rate of approximately 1,600 W, for a total of 9.6×10^{12} W. It is clear, then, that we are not on the verge of running out of energy in an absolute sense. What is in question is whether we can find sources of energy that will meet society's needs in a socially, environmentally, and economically acceptable way for the coming century.

If a healthy adult rides an exercise bike that drives an electrical generator, and the generator is hooked to a lightbulb, the best an average person can do with nonstop pedaling is to keep a 75-W bulb burning. Our ancestors realized this a long time ago, and they found ways to supplement human muscle power. First, they domesticated beasts of burden such as horses, oxen, camels, elephants, and llamas. Then they learned to make sails and use wind power, build dams to use waterpower, and convert the heat energy of wood, coal, and oil into mechanical power. Supplementary energy is now used in every part of our lives, from food production to transportation, manufacturing, housing, and recreation. An average North American uses energy, directly or indirectly, at a rate of 10,000 W—equivalent to 133 75-W lightbulbs burning continuously. (Note that a watt is, strictly speaking, a unit of power rather than energy; it refers to an amount of energy used per unit of time. Specifically, 1 W = 1 joule per second [J/s]. The same thing can be expressed in terms of calories per minute [1 watt = 14.34 cal/min], or British thermal units per hour [1 watt = 3.4129 Btu/hr]. You can find more information about units and conversions in Appendix 1.) North Americans are the world's biggest energy users; the worldwide average rate of energy use is only 1,600 W per capita because developing countries have much lower energy use.

Energy sources can be renewable or nonrenewable. Examples of renewable energy sources are fuelwood, wind, and waterpower for hydroelectricity. Nonrenewable energy sources include coal, oil, natural gas, and nuclear power. Everywhere in the world, even in the least developed countries, nonrenewable sources supply at least half of the energy used. The main source of nonrenewable energy used today comes from **fossil fuels**—coal, oil, and natural gas.

Throughout modern history, levels of energy consumption and the character of the world's energy "mix" have changed (Figure 11.3). During the Industrial Revolution, coal first began to replace wood as the main energy source in industrializing societies. Coal was burned to heat water, creating steam to drive the newly invented steam engines that powered the Industrial Revolution. After World War II, oil rose to predominance in the world's energy mix. The shift from coal to oil was driven by new technologies, primarily the internal combustion engine, which required fuel in liquid form. The ability to discover new fossil fuel deposits to meet these increasing demands was greatly improved by the advent of new technologies for exploration, including seismic studies, satellite imagery, computer modeling, and digital recording methods.

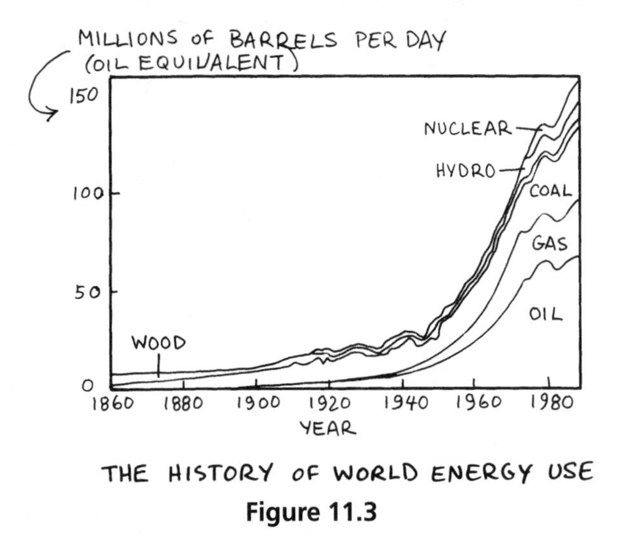

Figure 11.3

Despite our successes in finding new deposits, we cannot ignore the fact that fossil fuels are nonrenewable resources, and therefore finite. What happens when we reach the limit? We can assume (hopefully) that demand for energy will level off eventually, as a result of increased energy efficiency and conservation. Realistically, however, it is highly unlikely that the demand for fossil fuels will decrease substantially or that another fuel source will replace oil in the near future. Are supplies of fossil fuels adequate to meet demands through the next century? Comparing the estimated amounts of fossil fuels remaining, it appears that only coal, the most abundant (but also the most environmentally damaging) fossil fuel, may have the capacity to meet the world's demands beyond the next few decades. We must begin to look toward the day when alternative, renewable sources of energy will begin to fill a greater proportion of our energy needs.

What were the two new technologies that drove the shift from wood to coal and from coal to oil in the world's energy mix?

Answer: The steam engine and the internal combustion engine.

Fossil fuels consist of altered organic matter trapped in a sediment or sedimentary rock. The organic matter is the remains of either plants or animals, but various changes occur during and after burial. The kind of sediment, the kind of organic matter, and the kinds of postburial changes determine which type of fossil fuel will form.

Fossil fuels derive their energy from the Sun. Through photosynthesis, the Sun's energy is used to combine water (H_2O) and carbon dioxide (CO_2) into oxygen and organic carbohydrates (see chapter 10). When organic matter decays or burns, the solar energy trapped in the organic matter is released. Any organic matter that does not decay and is buried in sediment is a potential storage reservoir for solar energy. The total amount of organic matter trapped in sediments and sedimentary rocks is less than 1 percent of all the organic matter formed by plants and animals. Over geologic time, however, the absolute amount of trapped organic matter has grown to be very large. The chemical composition of this organic matter is dominated by the elements hydrogen and carbon, so an alternate name for fossil fuels is **hydrocarbons.**

Why are fossil fuels sometimes called hydrocarbons?

Answer: They consist of chemical compounds that are dominated by hydrogen and carbon.

5 PEAT AND COAL

Organic matter on land comes from trees, bushes, and grasses. Land plants are rich in organic compounds that tend to remain solid. In water-saturated places, such as swamps and bogs, the organic remains accumulate to form peat, an unconsolidated deposit of plant remains with a high carbon content (Figure 11.4). Peat is the initial stage of formation of the black, combustible rock called coal. Over millions of years, as layers of peat become compressed by overlying sediments, water is squeezed out and gaseous (volatile) compounds such as carbon dioxide (CO_2), and methane (CH_4) escape. By compaction and gas escape, peat is converted to lignite, bituminous coal, and eventually anthracite, the highest grade of coal. The conversion of peat (a sediment) to coal (a rock) is called **coalification.** Lower grades of coal are sedimentary rocks, but anthracite is so altered from its original state that it is a metamorphic rock. Coal occurs in strata (miners call them **seams**) along with other sedimentary rocks, mainly shales and sandstones. Most coal seams are 0.5 to 3.0 m (1.6 to 10.0 ft) thick, although some reach a thickness of more than 30 m (about 100 ft).

Peat has been formed more or less continuously since land plants first appeared. The size of swamps has varied greatly, however, and therefore the amount of peat and coal formed has also varied. By far the greatest period of coal swamp formation occurred during the Carboniferous and Permian periods. The great coal seams of Europe and eastern North America were formed at this time, when the swamp plants were giant ferns and scale trees (gymnosperms). The second great period of coal deposition peaked during the Cretaceous period. The plants of coal swamps during this period were flowering plants (angiosperms). Today, peat formation occurs in wetlands such as the Okeefenokee Swamp in

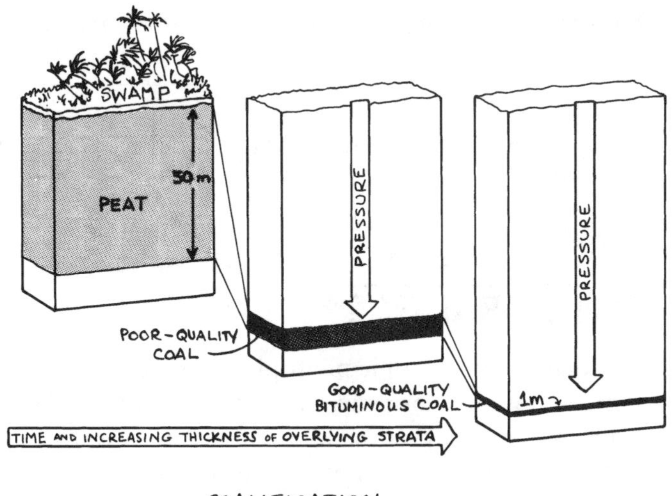

Figure 11.4

Florida and Georgia and the Great Dismal Swamp in Virginia and North Carolina. The Great Dismal Swamp, one of the largest modern peat swamps, contains an average thickness of 2 m (6.5 ft) of peat. However, unless this swamp lasts for millions of years, even that dense growth is insufficient to produce a coal seam as thick as some of the seams in Pennsylvania.

MINING COAL. Coal is being strip-mined here, near Kemmerer, Wyoming. The coal seam is the dark layer on which the shovel and trucks are standing. (Courtesy U.S. Department of Agriculture)

What are some examples of modern environments in which peat is forming today?

Answer: Peat is forming today in large swamps such as the Okeefenokee Swamp in Florida and Georgia and the Great Dismal Swamp in Virginia and North Carolina.

6 OIL AND NATURAL GAS

In the ocean, microscopic phytoplankton (tiny floating plants) and bacteria are the principal sources of organic matter trapped in sediment. Most of the organic matter is trapped in clay-rich sediment that is slowly converted to shale. During this conversion, organic compounds are transformed into oil and natural gas. These are the main forms of **petroleum**—naturally occurring gaseous, liquid, and semisolid substances that consist chiefly of hydrocarbon compounds. Petroleum (from the Latin words *petra,* "rock," and *oleum,* "oil") is referred to as **crude oil** when it emerges from the ground. From this state it must be distilled and refined.

Oil—the liquid form of petroleum—first came into use in 1847; by 1859, the first successful oil well was drilled in Titusville, Pennsylvania. **Natural gas**—naturally occurring hydrocarbons that are gaseous at ordinary temperatures and pressures—was first discovered in 1821; by 1872, it was being piped as far as 40 km (25 mi) from its source, for the purpose of lighting lamps. Today petroleum products are used for a very wide variety of purposes in addition to their main use as fuel for heating and running vehicles and machines. Different components of petroleum are broken down and used to make fertilizers, lubricants, asphalt, and an array of synthetic materials (such as rubber and plastic) and fabrics (such as nylon and rayon).

Deposits of petroleum are nearly always found in association with marine sedimentary rocks. When marine microorganisms die, their remains settle to the bottom and collect in the fine seafloor mud, where they start to decay. The decay process quickly uses up any oxygen that is present. The remaining organic material is preserved and covered with more layers of mud and decaying organisms. Over time, with continued burial, the muddy sediment with the partially decayed organic material is subjected to heat and pressure. This initiates a series of complex physical and chemical changes, called **maturation,** that break down the organic material and turn it into liquid and gaseous hydrocarbon compounds. Meanwhile, the muddy sediment itself turns into a rock. A rock containing organic material that is eventually converted into oil and natural gas is referred to as a **source rock.**

Oil and gas form in one rock formation and travel, or migrate, to another rock formation at a later time. The gas and small droplets of oil in the source rock are squeezed out by the weight of overlying rocks. For migration to occur, however, the oil must encounter a rock formation through which it can travel easily toward the surface, where the pressure is lower. Thus, migration requires rock with a high proportion of interconnected pore spaces, called a **reservoir rock** (Figure 11.5). Typically, the rocks with the highest porosity and permeability are sandstones and limestones, and these are the most common types of reservoir rocks.

Most of the petroleum that forms in sediments eventually makes its way to the surface and seeps out. It is not surprising, therefore, that the greatest proportion of existing oil and gas pools are found in rocks that are relatively young, no more than 2.5 million years old. This does not mean that older rocks produced less petroleum, but the oil in older rocks has had a longer time in which to escape. Sometimes an impermeable rock gets in the way of the migrating oil and gas and prevents it from going any farther. Such a formation is called a **cap rock.** The most common type of cap rock is shale. A geologic situation that includes a source rock to contribute organic material, a reservoir rock to allow for the migration of oil, and a cap rock to stop the migration is referred to as a **petroleum** (or **hydrocarbon**) **trap.**

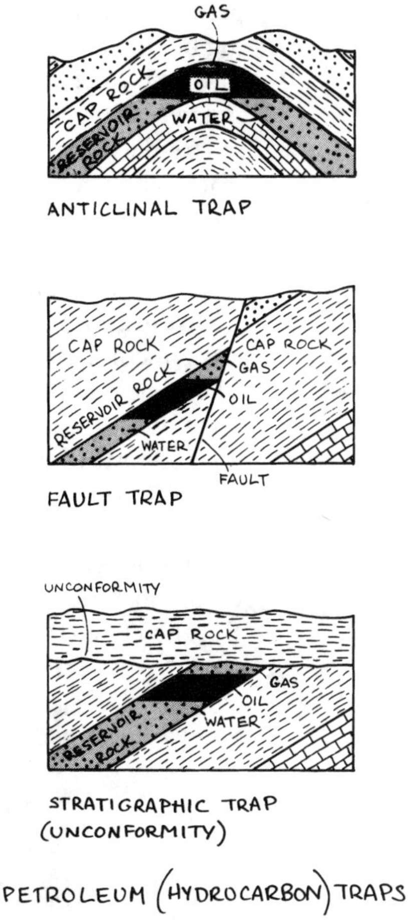

PETROLEUM (HYDROCARBON) TRAPS

Figure 11.5

What is petroleum?

Answer: Naturally occurring gaseous, liquid, and semisolid hydrocarbons, primarily oil and natural gas.

7 UNCONVENTIONAL HYDROCARBON RESOURCES

Hydrocarbon resources that are not oil, natural gas, or coal are called "unconventional" hydrocarbons. **Tar sands** are deposits of dense, viscous, asphaltlike oil that cannot be pumped easily. Tar is found in a variety of sedimentary rocks and unconsolidated sediments (not just sand, as the name implies). Tar sands may be petroleum deposits in which the volatile (light) fraction has migrated away, leaving behind a residual, tarry material. Alternatively, they may be immature deposits in which the chemical alterations that form liquid and gaseous hydrocarbons have not yet occurred. The largest known occurrence of tar sands is in Alberta, Canada, where the Athabasca Tar Sand covers an area of 5,000 km^2 (almost 2,000 mi^2) and reaches a thickness of 60 m (almost 200 ft). The Athabasca deposit may contain as much as 600 billion barrels of oil from tar (a barrel of oil is equal to 42 U.S. gallons). Similar deposits almost as large have been identified in Venezuela and in the former Soviet Union.

Kerogen is a waxlike organic compound that forms when organic material is buried, compacted, and cemented in very fine-grained sedimentary rocks such as lake shales. If burial temperatures are not high enough to initiate the chemical breakdowns leading to the formation of oil and natural gas, kerogen will be formed instead. If kerogen is heated, it breaks down and forms liquid and gaseous hydrocarbons similar to those found in oil and gas. All shales contain some kerogen, but to be considered an energy resource the kerogen in an **oil shale** must yield more energy than is required to mine and heat it. The world's largest deposit of oil shale is located in the United States. Millions of years ago, during the Eocene epoch, there were many large, shallow lakes in Colorado, Wyoming, and

OIL SHALE. The Green River Oil Shale in Colorado, Wyoming, and Utah contains the equivalent of 2,000 billion barrels of recoverable oil. The photo shows a canyon near Rifle, Colorado, where stream erosion has exposed strata of oil shale. (Courtesy Brian Skinner)

Utah. In three of them a series of organic-rich sediments was deposited, eventually forming the rock formations now known as the Green River Oil Shales. The Green River Oil Shale deposits could ultimately yield about 2,000 billion barrels of oil.

Where are the world's largest deposits of tar sands and oil shales?

____

____

Answer: The Athabasca Tar Sand deposits in Alberta, Canada, and the Green River Oil Shales in Colorado, Wyoming, and Utah.

8 ALTERNATIVE ENERGY SOURCES

Solar, biomass, wind, wave, tidal, hydroelectric, nuclear, and geothermal power are the major alternatives to fossil fuels. Some of these energy sources (biomass, in particular) are renewable. Others (nuclear power, in particular) are technically nonrenewable, but the amount of energy available is so large that they are effectively inexhaustible. Tidal energy, solar energy, and other sources that derive their energy from the Sun (such as wind and waves) are also inexhaustible—at least as long as the Sun-Moon-Earth system exists.

Solar energy reaches the Earth from the Sun at a rate more than 10,000 times greater than the energy humans use from all other sources combined. Direct solar energy is best suited to supplying heat for such applications as home heating and water heating for home use. Converting solar energy into electricity is a major technological challenge. Photovoltaic cells that effect this conversion have been invented, but so far their cost is too high and their efficiency too low for most uses; they are mainly used in small solar-powered calculators, radios, and the like. Photovoltaic technology is constantly improving, however, and the cost of energy generated in this manner is decreasing.

Solar energy can be stored for long periods of time in the form of fossil fuels, but recently living plant matter also contains stored solar energy. Any form of energy that is derived more or less directly from the Earth's plant life is **biomass energy.** In the form of wood, biomass was the dominant source of energy until the end of the nineteenth century, when it was displaced by coal. Biomass fuels—primarily wood, peat, animal dung, and agricultural wastes—are still widely used throughout the world, especially in developing countries. Organic materials such as animal dung can be converted into methane for power generation; this form of fuel is called **biogas.**

Winds are a secondary expression of solar energy. For thousands of years, wind has been used as a source of power for ships and windmills. Today, huge windmill "farms" are being erected in suitably windy places. In Denmark, about 3,000 wind turbines supply electricity throughout the country. Although there are some problems with windmill technologies, it seems very likely that windmills will soon be cost competitive with coal-burning electrical power plants. Unfortunately, steady surface winds can provide only about 10 percent of the amount of energy now used by humans. Therefore, wind power may be locally important but probably will not become globally significant.

Waves are created by winds; they are another expression of solar energy. Waves contain an enormous amount of energy, but so far no one has discovered how to tap them as a source of power on a large scale. Tides are another ocean-related power source. Their energy comes from the rotation of the Earth and gravitational interaction with the Moon and the Sun (see chapter 9). If a dam is constructed across the mouth of a narrow bay so that water is trapped at high tide, the water can be released at low tide to drive a turbine. The use of tidal energy is not new; a mill in Britain dating from 1170 still runs on tidal power. Like wind power, tidal power is insufficient to satisfy more than a small fraction of human energy needs and thus can only be locally important.

Hydroelectric energy is the only form of water-derived power that currently fulfills a significant portion of the world's energy needs. Hydroelectric power is generated from the energy of a flowing stream of water; it is primarily gravitational energy. To convert the power of flowing water into electricity, it is necessary to build dams. The flowing water is used to run turbines, which convert the energy into electricity. The total recoverable energy from the water flowing in all of the world's streams has been estimated to be equivalent to the energy obtained by burning 15 billion barrels of oil per year. Thus, even if *all* the potential hydropower in the world were developed, it could not satisfy today's energy needs (currently the equivalent of about 30 billion barrels of oil per year). Another problem is that there are many negative environmental side effects associated with the development of large hydroelectric dams. Reservoirs eventually fill with silt, so even though waterpower is continuous, dams and reservoirs have limited lifetimes.

Geothermal energy, the Earth's internal energy, is used commercially in a number of countries, including New Zealand, Italy, Iceland, and the United States. The most easily exploited geothermal deposits are hydrothermal reservoirs, underground systems of circulating hot water and/or steam in fractured or porous rocks near the surface. The reservoirs are near the surface because the sources of heat are shallow magma chambers or thick piles of recently erupted volcanic rocks. An example is The Geysers in northern California—the largest producer of geothermal power in the world. To be used efficiently for geothermal power, hydrothermal reservoirs should be 200°C (almost 400°F) or hotter, and this temperature must be reached within 3 km (less than 2 mi) of the surface. Most of the world's hydrothermal reservoirs are close to the margins of tectonic plates where recent volcanic activity has occurred and hot rocks or magma are close to the surface.

Give examples of nonrenewable, renewable, and inexhaustible energy sources.

Answer: Biomass energy sources (wood, animal dung, agricultural waste, biogas, etc.) are renewable. Inexhaustible sources include tidal, wind, solar, wave, geothermal, and hydroelectric energy. Nonrenewable energy sources include fossil fuels and nuclear energy (the latter nonrenewable but essentially inexhaustible).

9 NUCLEAR ENERGY

Nuclear energy is the heat energy produced during the controlled transformation of an isotope of one chemical element to an isotope of another chemical element. Nuclear energy can be generated in two ways: by splitting heavy atoms into lighter atoms, a process called **fission,** or by combining two light atoms to make a heavier atom, a process called **fusion.**

Three of the radioactive atoms that keep the Earth hot by spontaneous radioactive decay—uranium-235, uranium-238, and thorium-232—can be mined and used to obtain nuclear energy by fission. When a fissionable atom is hit by a neutron it not only releases heat and forms new, lighter elements, it also ejects some neutrons from its nucleus (Figure 11.6). These neutrons can then be used to induce more atoms to fission, creating a continuous chain reaction. If the operation is carried out in a device called a pile, which can control the rate of neutron bombardment, the rate of fission can be controlled and the pile becomes a heat source. (When a chain reaction proceeds without control, an atomic explosion occurs.) The fissioning of

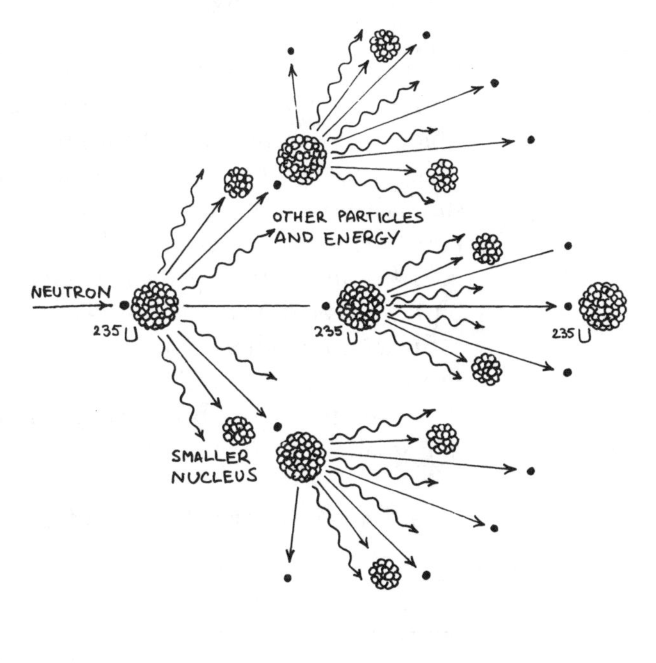

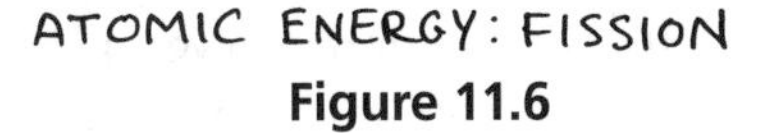
ATOMIC ENERGY: FISSION

Figure 11.6

just 1 gram of uranium-235 produces as much heat as the burning of 13.7 barrels of oil.

Nuclear power plants use the heat energy from fission to produce steam that drives turbines and generates electricity. Approximately 17 percent of the world's electricity is derived from nuclear power plants. In France, more than half of all the electrical power comes from nuclear plants; the proportion is rising sharply in some other European countries and in Japan. The reason for the increase is that Japan and most European countries do not have adequate supplies of fossil fuels to be self-sufficient. Nuclear power is considered—even by some environmentalists—to be a clean source of energy, because it has no harmful atmospheric emissions. The main problem plaguing the nuclear power industry is that fission generates highly radioactive by-products—high-level nuclear waste—which must be isolated from the biosphere and hydrosphere for many thousands of years. This presents a technically and socially difficult disposal problem that has not yet been resolved.

In principle, nuclear fusion—the joining together or fusing of two small atoms to create a single larger atom, with an attendant release of heat energy—is another potential source of nuclear power. Nuclear fusion utilizes as its fuel a heavy isotope of hydrogen, called deuterium. The Earth has a virtually endless supply of deuterium in the form of a very common chemical compound—water (hydrogen dioxide, or H_2O). The primary by-product of nuclear fusion would be helium, a nontoxic, chemically inert gas.

This conjures up images of a cheap, clean power source with a virtually endless fuel supply. So why are we not using energy provided by nuclear fusion? Fusion is the nuclear process that occurs in the cores of stars, the process responsible for the tremendous heat energy generated by the Sun. But that, in a nutshell, is the problem. For two nuclei to fuse, the ambient conditions must be similar to those at the core of a star—on the order of 100 million degrees. The possibility that nuclei could be induced to fuse at room temperature (so-called cold fusion), bringing instant scientific recognition to those who first accomplish it, has led many scientists to search for this "holy grail" of energy. In 1989, a furor arose when two researchers announced that they had achieved cold fusion in a test tube. Unfortunately, the work of those researchers was not substantiated in other laboratories. The routine use of fusion power remains a distant goal.

What is the difference between fission and fusion?

____

____

Answer: Fission involves splitting a heavy atom into lighter atoms. Fusion involves combining two light atoms to make a heavier atom.

SELF-TEST

These questions are designed to help you assess how well you have learned the concepts presented in chapter 11. The answers are given at the end.

1. "Source rock" refers to rocks ___________.
   a. in which organic matter accumulated
   b. from which oil and gas are drawn
   c. that trap oil and gas below the ground
   d. All of the above are true.
2. The nonvaluable minerals associated with ores are called ___________.
   a. bauxite
   b. gangue
   c. placers
   d. kerogen
3. The world's largest known occurrence of tar sand is in ___________.
   a. the United States
   b. Canada
   c. Russia
   d. Iceland
4. The Earth's energy comes from three sources. They are ___________, ___________, and ___________.
5. ___________ is a waxlike organic substance commonly found in sedimentary rocks.
6. Iron-rich laterites and aluminum-rich bauxites are examples of ___________ mineral deposits.
7. All ores are mineral deposits, but all mineral deposits are not ores. (T or F)
8. Shale and limestone are common source rocks for oil and natural gas. (T or F)
9. Wind and wave energy are both expressions of solar energy. (T or F)
10. The only way that diamonds are known to reach the surface of the Earth is through kimberlite pipes. (T or F)
11. What is the difference between fossil fuels, petroleum, and hydrocarbons?

____

____

____

____

12. In what tectonic environments is exploitable geothermal energy most likely to be located? Give three examples of places where geothermal energy is now exploited commercially.

____

____

____

ANSWERS

1. a
2. b
3. b
4. solar radiation; geothermal energy; tidal energy (or) the Sun; the Earth's internal energy; tidal interactions among the Earth, Sun, and Moon
5. kerogen
6. residual (or) weathering-related
7. T
8. T
9. T
10. T
11. Fossil fuels are the altered remains of organic matter trapped in sediment or sedimentary rock. The principal fossil fuels are coal, oil, and natural gas. Petroleum includes oil and natural gas (naturally occurring liquid, gaseous, and semisolid hydrocarbons). Hydrocarbons are compounds that consist principally of hydrogen and carbon.
12. Close to the margins of tectonic plates, where volcanic activity has occurred recently, and where hot rocks or magma can be found close to the surface. Examples discussed in the text are New Zealand, Italy, United States (California), and Iceland.

KEY WORDS

biogas
biomass energy
cap rock
coalification
crude oil
deposit
economic geology
exploration geology
fission
fossil fuel
fusion
gangue
geothermal energy
grade

hydrocarbon
hydrothermal solution
kerogen
maturation
metallic mineral resource
native metal
natural gas
nonmetallic mineral resource
nonrenewable resource
nuclear energy
oil
oil shale
ore
petroleum
petroleum (hydrocarbon) trap
placer deposit
renewable resource
reservoir rock
residual deposit
seam (coal)
source rock
tar sand

12 Earth Systems and Cycles

And some rin up hill and down dale, knapping the chucky stanes to pieces wi' hammers, like sae mony road makers ron daft. They say 'tis to see how the world is made.

—Sir Walter Scott

The Earth has a spirit of growth.

—Leonardo da Vinci

OBJECTIVES

In this chapter you will learn

- how scientists approach the study of the Earth as a unified system;
- how the tectonic cycle, the rock cycle, and the hydrologic cycle interact to shape the world we live in;
- how internal and external geologic processes are linked;
- how plate tectonics provides a global context for our observations about rock-forming processes.

1 EARTH SYSTEM SCIENCE

The traditional way of studying the Earth has been to focus on separate units—the atmosphere, the ocean, a mountain range, a lake—in isolation from the others. It is still important to learn about the individual parts of the Earth, but now scientists are also interested in studying the interactions and interrelationships among the various parts of the Earth system. Increasingly, the Earth is being studied as a whole and viewed as a unified system. Earth system science helps geologists break down a large, complex problem into smaller pieces that are easier to study, without losing sight of the interconnections among those pieces.

A **system** is any portion of the universe that can be separated from the rest of the universe for the purpose of observing changes. By saying that a system is "any portion of the universe," we mean that the system can be whatever the observer defines it to be. A system can be large or small, simple or complex. You could choose to observe the contents of a beaker in a laboratory experiment. Or you might study a fist-sized sample of rock, an ocean, a volcano, a mountain range, a continent, a planet, or the entire solar system. All of these are examples of systems; some are more complex than others. A leaf is a system, but it is also part of a larger system (a tree), which in turn is part of an even larger system (a forest), and so on.

The fact that a system is "separated from the rest of the universe for the purpose of observing changes" means that it has boundaries that set it apart from its surroundings. The nature of those boundaries is one of the most important characteristics of a system, leading to three basic kinds of systems (Figure 12.1). In an **isolated system,** the boundaries are such that they prevent the system from exchanging either matter or energy with its surroundings. The concept of an isolated system seems easy to understand, but such a system could only be imaginary. In the real world, it is possible to have boundaries that prevent the passage of matter, but it is impossible for any boundary to be so perfectly insulating that energy can neither enter nor leave the system.

A second type of system is a **closed system,** which has boundaries that permit the exchange of energy, but not matter, with its surroundings. An example of a closed system would be a perfectly sealed oven, which would allow the material inside to be heated but would not allow any of that material to escape. (In real life, ovens allow both heat and food vapors to escape, so they are not perfect examples of closed systems.) The third kind of system, an **open system,** can exchange both matter and energy across its boundaries. An island is an example of an open system (Figure 12.2). Matter (in the form of water) enters the system as precipitation

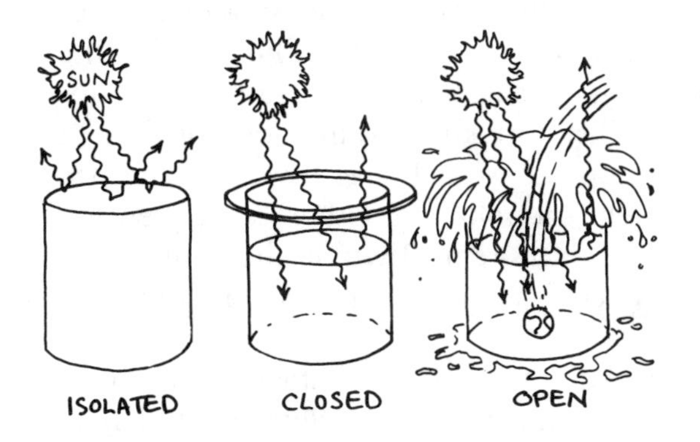

THREE TYPES OF SYSTEMS

Figure 12.1

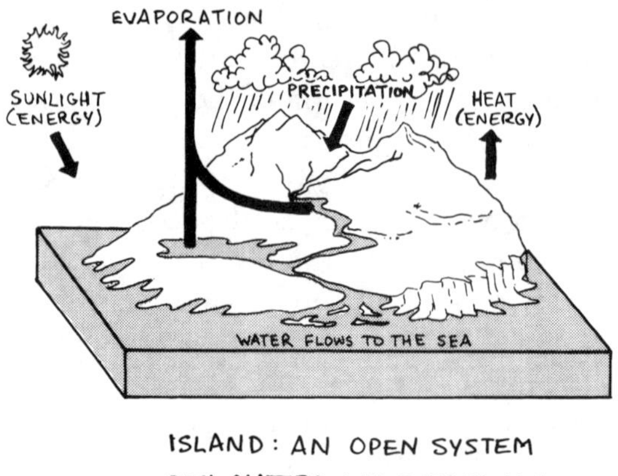

ISLAND: AN OPEN SYSTEM
BOTH MATTER AND ENERGY CAN ENTER AND LEAVE THE SYSTEM.

Figure 12.2

and leaves by flowing into the sea or by evaporating back into the atmosphere. Energy enters the system as sunlight and leaves as heat.

Geologists use system science to study the Earth and the relationships among the atmosphere, the biosphere, the hydrosphere, and the lithosphere. Earth itself approximates a closed system. Energy enters the Earth system as solar radiation; the energy is used in various biologic and geologic processes, then leaves the system as heat. It is not quite correct to say that no matter crosses the boundaries of the Earth system; we lose some hydrogen and helium atoms from the outermost part of the atmosphere, and we gain some material in the form of meteorites. However, the amount of matter that enters or leaves the planet on a daily basis is so minuscule compared with its overall mass that the Earth essentially functions as a closed system.

When a change is made in one part of a closed system, it affects other parts of the system. The various subsystems of the Earth are in a dynamic state of balance. When something disturbs one subsystem, the others also change as the balance or equilibrium is reestablished. The fact that the Earth is a closed system has other important implications. By definition, the matter in a closed system is fixed and finite. Therefore, the resources on this planet are all we have and, for the foreseeable future, all we will ever have. We must treat Earth resources with respect and use them cautiously. Another consequence of living in a closed system is that waste materials remain within the boundaries of the system. As environmentalists say, "There is no *away* to throw things to."

Think about natural environmental systems, such as trees, forests, groundwater aquifers, lakes, and so on. Which kind of system do you think is most common in the natural environment: closed or open systems?

Answer: Most natural environmental systems are open systems—both materials and energy cross their boundaries.

2 CYCLES

It is useful to envision interactions within the Earth system as a series of interrelated **cycles,** groups of processes whereby materials and energy move among the Earth's reservoirs. Earth system science helps us study how materials and energy are stored and how they are cycled among the four principal reservoirs of the Earth system—the atmosphere, the biosphere, the hydrosphere, and the lithosphere. Because this book is about physical geology, it has focused mainly on processes in the lithosphere. However, the systems approach tells us that it is unreasonable—even impossible—to consider one part of the Earth system in isolation from the rest. Physical geology deals with processes in the lithosphere and the interior of the Earth, but these are intimately associated with processes in the hydrosphere, the atmosphere, and the biosphere. In this book, you have not only learned geology but some oceanography, hydrology, meteorology, physics, chemistry, biology, and astronomy as well.

Using the concept of cycles, we can trace the movement of energy and materials from one reservoir to another (Figure 12.3). The cycles that are particularly important in physical geology are:

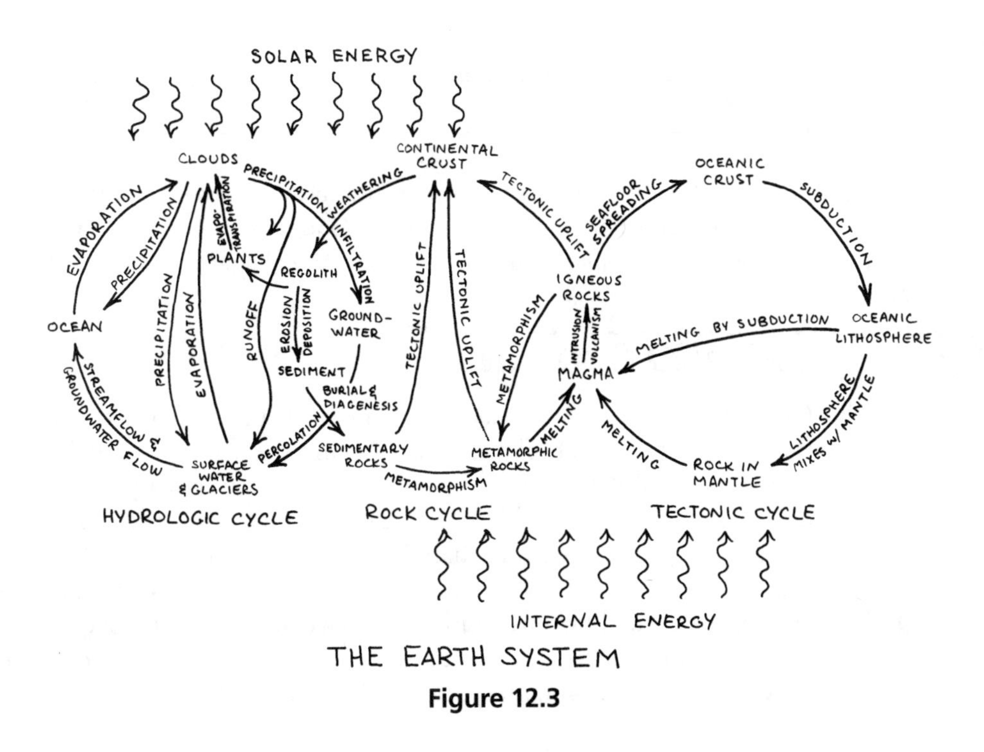

Figure 12.3

- The **tectonic cycle,** which deals with the movements and interactions of lithospheric plates and the internal processes that drive plate motion
- The **rock cycle,** which describes all of the various crustal processes by which rock is formed, modified, transported, broken down, and re-formed
- The **hydrologic cycle,** which describes the movement of water from reservoir to reservoir within the Earth system

These cycles are linked together through the various geological, physical, chemical, and biological processes you have learned about in this book. The processes are driven by energy that comes from external sources (mainly the Sun) and from the Earth's own internal energy. When rocks are uplifted and exposed as a consequence of plate tectonics, they immediately start to weather and break down. Like two giant sculptors working on the same statue—one from the inside, the other from the outside—plate tectonics and weathering constantly change and reshape the Earth's surface. We observe that the kinds of rocks forming today are the same as the rocks we can see in ancient continents. From this we conclude that the tectonic cycle, the hydrologic cycle, and the rock cycle have been operating in more or less the same way for billions of years.

What are the main energy sources that drive the tectonic cycle, the rock cycle, and the hydrologic cycle?

____

____

____

Answer: The tectonic cycle is driven by the Earth's internal (geothermal) energy. The hydrologic cycle is driven by energy from the Sun (and, to a lesser extent, by gravity). The rock cycle is driven by a combination of internal and external energy sources.

3 CRATONS AND OROGENS

Geologic evidence suggests that plate tectonics has been the dominant Earth-shaping process for at least the past 2 billion years, and probably longer. Evidence of present-day plate tectonics comes mainly from oceanic crust, but all of today's oceanic crust is young—less than 200 million years old. Therefore, evidence of plate tectonic activity more than 200 million years old must be sought in the older rocks of the continental crust. Part of this evidence is found in the way continents are constructed. They are made up of two structural units: cratons and orogens.

A **craton** is a core of ancient continental crust that is tectonically and isostatically stable. Tectonic stability means that the crust has not been involved in a plate collision for at least the last billion years. Isostatic stability arises from isostasy, the flotational balance of the lithosphere on top of the asthenosphere (defined in chapter 1). "Floating" is not exactly the right word, but we use it because it adequately describes what happens to lithospheric plates on top of the asthenosphere. The asthenosphere, even though solid, is hot and weak; it is easily deformed in comparison to the rigid overlying lithosphere. Lithospheric plates "float" on the asthenosphere, like a block of wood floats on water (Figure 12.4). In response to tectonic deformation or changes in mass, a block of lithosphere may sink or rise, just like the block of wood will sink when a weight is placed on it, or rise when the weight is removed. When a craton reaches isostatic stability, there is no tendency for the crust to rise or sink—it is like a block of wood floating on a still pond.

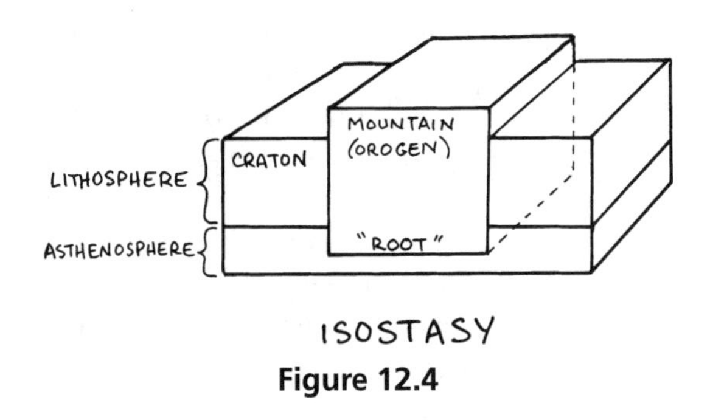

Figure 12.4

An example of part of a craton that is in a temporary state of isostatic instability is the North American Craton in the area of the Great Lakes (technically called Superior Craton). During the last ice age, this area was weighed down under a sheet of ice up to 2.5 km (1.5 mi) thick. The crust sagged under the weight of this great thickness of ice. After the ice melted and receded at the end of the Pleistocene glaciation (leaving behind the Great Lakes), the crust began to rebound. However, a huge block of lithosphere can't just pop up like a block of wood in water. The rocks of the craton itself and the other rocks that surround it have a great deal of strength, unlike water, and they prevent the craton from rising quickly. Consequently, the crust under the Great Lakes is still rising—very slowly—in response to the removal of the weight of all that ice. This slow, isostatic uplift is thought to have been responsible for a number of earthquakes in this part of North America.

The other type of structural unit of which continents are constructed is called an orogen. **Orogens** are long, thin portions of crust that have been intensely deformed and metamorphosed during a continental collision—they are mountain belts, or the ancient remnants of them. (The terms "orogen" and "orogenesis," which means "mountain building," come from the Greek words *oros,* "mountain," and *genesis,* "to come into being.") The Himalayas are an example of a present-day orogen; they are not tectonically stable because they are in an active collision zone, where the crust is still being deformed and uplifted to form a great mountain chain. Through radiometric dating, some orogens have been found to be as old as 3.8 billion years. This provides geologists with evidence confirming the antiquity of lithospheric plates, plate collisions, and the processes of mountain building. Ancient orogens, now deeply eroded, betray their history through a combination of deformational and metamorphic features.

Many orogens, both ancient and modern, are not isostatically stable. Why is this so? Beneath every great mountain range there is a deep root of thickened continental crust, poking down into the asthenosphere. Thus, a mountain range is like an iceberg floating in the sea. As erosion weathers away the top of a mountain range, like sunshine melting an iceberg, the root at the base of the mountain range slowly rises to maintain its flotational—that is, isostatic—balance. This is why mountain ranges like the Appalachians, which are no longer tectonically active, can still have topographic relief even after millions of years of erosion.

An assemblage of cratons and orogens that has reached isostatic stability is called a **continental shield.** For example, North America has a huge continental shield at its core. It consists of several ancient cratons and orogens (mostly over 2 billion years old), surrounded by four younger orogens: the Applachian, Cordilleran, Innuitian, and Caledonide mountain belts (Figure 12.5). The rocks of this continental shield outcrop at the surface in Canada but are mostly covered by thick layers of sedimentary rocks in the United States; for this reason, the North American continental shield is usually called the Canadian Shield.

Let's summarize. Ancient orogens establish the antiquity of plates, plate collisions, and mountain-building processes. The isostatic stability of ancient cratons proves that isostasy, too, has operated for an exceedingly long time. If the litho-

ANCIENT OROGEN. The Appalachian Mountains of central Pennsylvania are the eroded remains of a 300-million-year-old orogen formed in the Paleozoic era. Forested ridges are underlain by rocks such as sandstone and quartzite that resist erosion. Soft, easily eroded rocks such as shale and limestone underlie the valleys. (Earth Satellite Corporation)

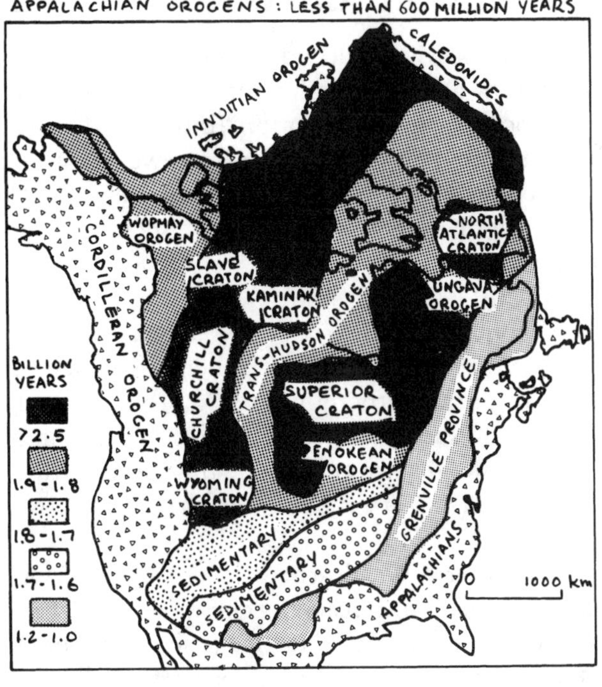

CRATONS AND OROGENS:
THE NORTH AMERICAN CONTINENTAL SHIELD

Figure 12.5

sphere did not float isostatically on the asthenosphere, and if the lithosphere did not break into plates, plate tectonics could not happen. The formation of most kinds of igneous, metamorphic, and sedimentary rocks is controlled directly or indirectly by plate tectonic processes. Therefore, the distribution of most kinds of rocks on the Earth can best be understood in terms of plate tectonics.

It's worth reemphasizing the point that plate tectonics is the framework that unifies what we know about rocks and rock formation. Instead of dealing with a fragmentary collection of observations about local occurrences and processes, geologists now understand and interpret these observations in the larger context of the global tectonic cycle. In this final chapter of the book, we will revisit some of the most important rock types and consider what our understanding of global tectonics can tell us about rock-forming processes—and vice versa.

Is there such a thing as a modern craton?

Answer: No; by definition, cratons are ancient.

4 WHERE DIFFERENT IGNEOUS ROCKS FORM AND WHY

Igneous rock is crystallized magma. Therefore, the distribution of igneous rock reflects how and where different magmas form. Let's briefly review a few key concepts about magma formation (introduced in chapter 5). These basic concepts about magma will come in handy as we examine different types of igneous rocks in the context of the tectonic environments in which they occur.

Recall that magma forms by the melting of rock and that melting can happen in three ways:

1. When temperature rises
2. When pressure drops (in general, the melting temperatures of rocks increase with pressure; therefore, when a hot mass of rock is under pressure and that pressure suddenly decreases, decompression melting may occur)
3. When water lowers the melting temperature of the rock ("wet melting," as opposed to "dry melting" in the absence of water)

In chapter 5 you learned that partial melting also plays an important role in magma formation. Because rocks are assemblages of minerals, they do not melt at a single temperature, the way ice (a single mineral) melts to water. Rocks melt over a temperature range that may span several hundred degrees. The ratio of melted rock (magma) to unmelted rock slowly changes across the melting interval, and the composition of the magma also changes. Magma composition, therefore, is determined by the extent of melting; for a given starting rock, the composition of the melt at 5 percent partial melting will be different from the composition of the melt at 10 or 20 or 50 percent partial melting, for example.

Let's look first at igneous rocks that form in oceanic settings. We will consider three important types of oceanic igneous rocks: midocean ridge basalt (MORB), ophiolites, and volcanic rocks that form over oceanic mantle "hot spots." Next we will look at igneous rocks that form in subduction zones, where two lithospheric plates collide. Then we will turn to igneous rocks that form in a variety of continental settings.

What are the three conditions that can lead to rock melting?

Answer: Increase in temperature, decrease in pressure, and presence of water.

5 IGNEOUS ROCKS IN OCEANIC ENVIRONMENTS

Ocean basins are everywhere underlain by basaltic crust. New oceanic crust is formed at midocean ridge spreading centers, large fractures along which the lithosphere is rifting, or splitting apart. Beneath a midocean ridge, the hot rock of the asthenosphere moves upward, squeezing into the fracture zone. As it does this, decompression melting of the asthenosphere produces basaltic magma (Figure 12.6). The magma that solidifies to form oceanic crust varies little in composition around the world. It is referred to as **MORB,** which is the acronym for **mid**o**c**ean **r**idge **b**asalt.

Even though it is the same worldwide, MORB can reveal much to geologists about the processes by which oceans and oceanic crust are formed. The decompression melting that produces MORB magma starts in the mantle when the upwardly moving rocks of the asthenosphere reach a depth of about 65 km (40 mi). A 5 percent partial melt of mantle rock produces a magma of MORB composition. If a larger or smaller fraction of the mantle rock were to melt, the magma would not be of MORB composition. The reason that all MORB is similar is because of the similarity of the tectonic environment in which these magmas form worldwide—the depth of decompression melting and the degree of partial melting vary little from one midocean ridge to another.

Why is MORB more or less the same everywhere in the world?

Answer: Because of the similarity of the tectonic environments in which all MORB magmas are formed.

Ocean basins cover 71 percent of the Earth's surface; consequently, MORB is the most common igneous rock at the Earth's surface. However, MORB is not easy to observe. With the exception of a few rare places such as Iceland, where the midocean ridge itself actually pokes up above sea level, the ridge and the seafloor are covered everywhere by water. There is, however, one special cir-

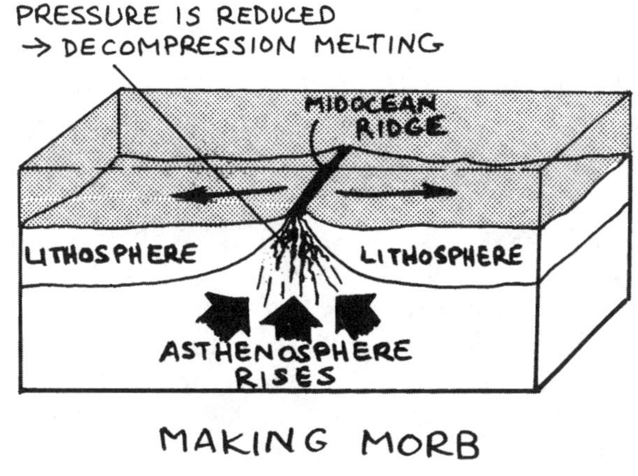

Figure 12.6

MIDOCEAN RIDGE. Thingvellir Graben, which crosses the southwest corner of Iceland, is one of the few places in the world where an active midocean spreading ridge can be seen above water. (Courtesy S. Thorarinsson)

cumstance where it is possible to see MORB above sea level: where a plate collision has caught and crushed a fragment of oceanic crust between two colliding continents. When MORB is found on land, it usually occurs as the fragmented remnants of such a collision. In this process, the basalt becomes metamorphosed to a distinctive mineral assemblage called **spilite,** dominated by the fibrous mineral serpentine. These green, serpentine-dominated fragments of oceanic crust, along with the rocks that characteristically occur with them, are called **ophiolites** (from the Greek word for "serpent," *ophis*). Ophiolites are studied intensively by geologists because they provide evidence of ancient plate collisions. They also provide convenient samples of oceanic crust, which is otherwise difficult to study.

A cross section through an ophiolite is essentially a cross section through a typical segment of oceanic crust (Figure 12.7). At the top is a thin veneer of ocean-floor sediments. Beneath the sediments are layers of "pillowed" basaltic lavas; the round, pillowlike structures indicate that the magma was extruded underwater. Still deeper are sills made of gabbro (the plutonic equivalent of basalt). Cutting through the sills and pillow lavas are many vertical dikes of gabbro. The basalts and gabbros have the same composition—MORB—and clearly originated from the same magma. The gabbro dikes and sills were the channels

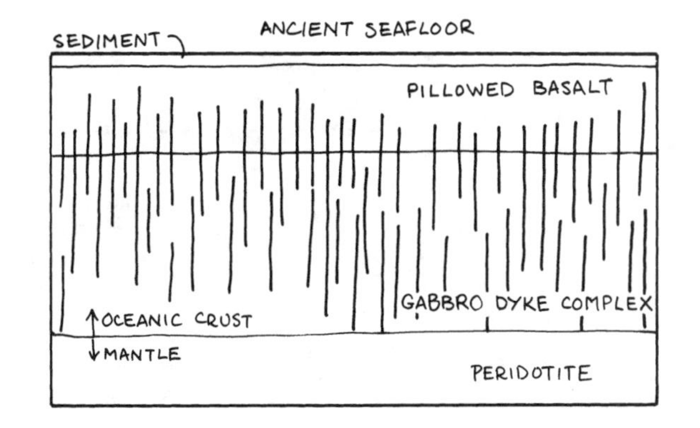

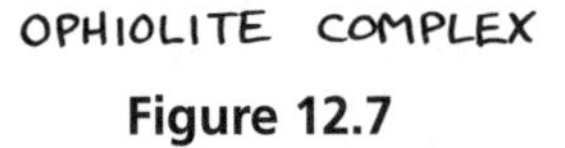

Figure 12.7

through which the magma passed as it ascended from the upper mantle to the ocean floor. In some ophiolites, rock of a very different composition from MORB is found beneath the gabbro dikes and sills. It is peridotite, an olivine- and pyroxene-bearing rock that is characteristic of the mantle. Ophiolites thus provide samples of ancient oceanic crust and, in some cases, the upper mantle. The contact between the gabbro and the peridotite is, of course, a sample of the Moho—the crust-mantle boundary.

MORB is the most common igneous rock on the surface of the Earth; why is it so difficult to study?

Answer: MORB forms the ocean floors, so it is mostly covered by deep water.

On the world's ocean floors there are also many shield volcanoes, some of which rise above sea level to form volcanic islands. The Hawaiian Islands are probably the best known, but there are many others. Many features make these shield volcanoes distinctive, but two are especially important. First, the composition of the magma is similar to MORB and so probably has a similar origin. The second distinctive feature is that each of today's active oceanic shield volcanoes—such as Kilauea in Hawaii—is the youngest in a chain of progressively older volcanoes.

Why are these features significant? The magmas that form shield volcanoes like Kilauea are similar to MORB but not exactly the same, and the differences can't be fully explained by differing degrees of partial melting. Geochemical experiments show that basaltic shield volcano magmas are formed by partial melting in the mantle, but at higher pressure—and therefore deeper in the mantle—than MORB magmas. A chain of volcanoes all extruding the same magma implies

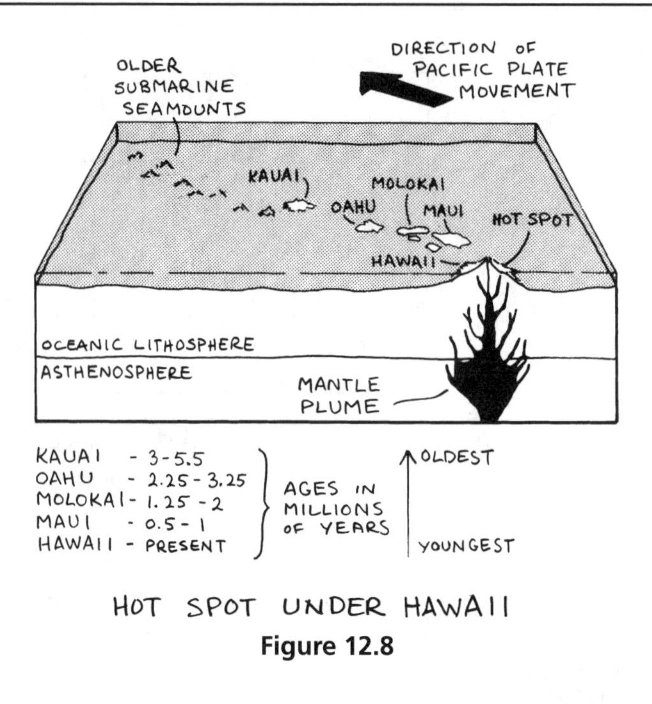

HOT SPOT UNDER HAWAII

Figure 12.8

a long-lived, stationary, deep-seated source—a **hot spot.** The mantle hot spot feeds magma to the active volcano on top of it (Figure 12.8). Meanwhile, the lithospheric plate drifts along, carrying the volcano with it. Eventually the old volcano moves off the hot spot altogether; without an underlying magma source, the volcano becomes inactive. A new, active volcano is "born" on top of the hot spot, and a line of progressively older volcanoes records the history of plate motion over the hot spot.

Where does the magma come from that feeds these hot-spot volcanoes? The most plausible explanation is that a long, thin stream of hot rock, called a **mantle plume,** rises from deep in the mantle, perhaps as deep as the core-mantle boundary. Reaching a depth of 100 to 200 km (about 60 to 120 mi) below the surface, the rocks in these plumes begin to undergo decompression melting. The rising plumes are part of the great mantle convection system discussed in chapter 1. The Martian volcano Olympus Mons, the largest volcano so far discovered in the solar system, is thought to have originated through a similar process. It is a basaltic shield volcano, similar to the Hawaiian volcanoes, but very much larger. The volcanic edifice of Olympus Mons probably grew from magma fed to the surface by a very long-lived plume deep within the planet.

What is the largest volcano in the solar system, and what type of volcano is it?

Answer: Olympus Mons (Mars); it is a basaltic shield volcano, fed by a mantle plume.

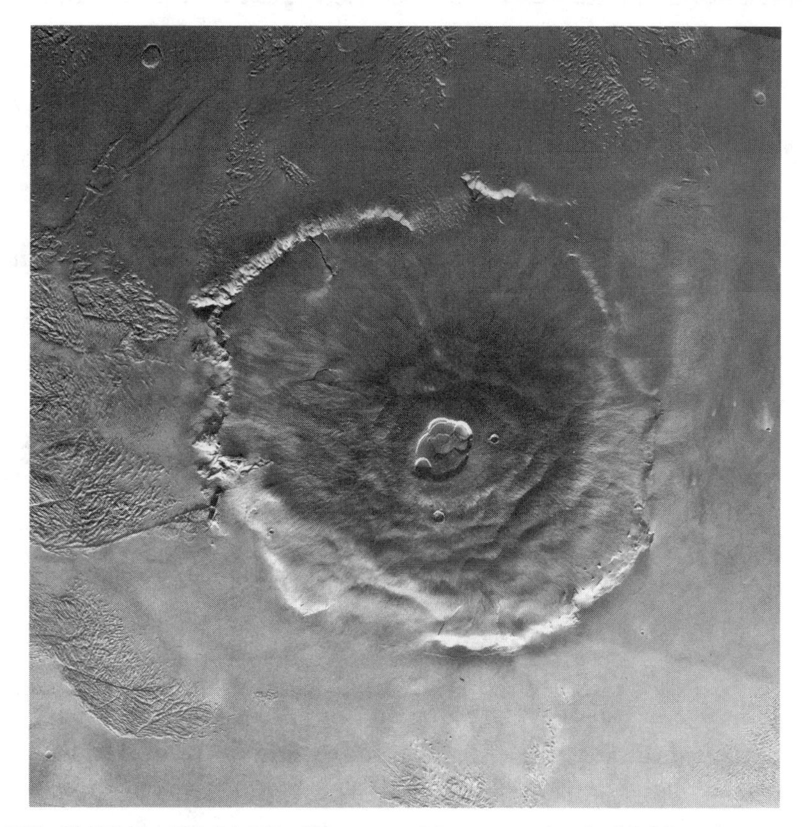

PLUME-FED SHIELD VOLCANO. Olympus Mons, a giant shield volcano on Mars, dwarfs all of Earth's volcanic structures. Mars apparently does not have plate tectonics, so a rising plume of magma can continue to build up a volcanic edifice for a very long time. On the Earth, as shown by the Hawaiian Islands, plate movement soon breaks the connection between a volcano and its plume-generated magma source. (Courtesy NASA)

6 IGNEOUS ROCKS IN SUBDUCTION ZONE ENVIRONMENTS

From the tectonic cycle, we know that after new oceanic crust forms at the mid-ocean ridges it spreads laterally, forming great ocean basins. Eventually, it will meet another fragment of crust—either oceanic or continental—along a convergent plate margin, and subduction will occur. This cycle of production of new oceanic crust, opening of the ocean basin, lateral movement of oceanic crust, and eventual consumption of the crust in a subduction zone is called a **Wilson cycle.** It is named for Canadian geophysicist J. Tuzo Wilson, who first proposed it in the 1960s as an explanation for the geologic history of the Appalachian Mountains. The special tectonic environment of active convergent plate margins gives rise to magmas that are more variable in composition than MORB.

Along an ocean-ocean convergent plate margin, oceanic lithosphere is subducted into the mantle. As the slab sinks, it heats up, and hydrous (water-bearing) minerals in the oceanic crust begin to release water. This water induces wet partial melting in the part of the mantle that is immediately adjacent to (and overlying) the subducting slab (Figure 12.9). Melting begins when the slab reaches a depth of about 100 km (60 mi). The subducting slab of oceanic crust itself may also undergo a small amount of wet partial melting. Magma that forms as a result of this type of process is andesitic in composition.

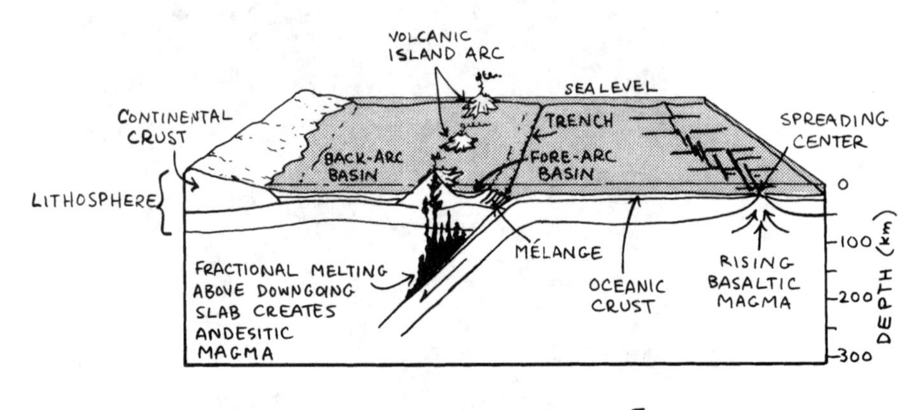

Figure 12.9

The andesitic magma that forms in subduction zones rises through the overlying mantle and erupts as explosive stratovolcanoes. These volcanoes occur as chains of islands—called **volcanic island arcs**—parallel to the deep ocean trench that marks the edge of the subducting plate (Figure 12.9). The Aleutians, Japan, the Philippines, and the Kamchatka Peninsula are examples of modern volcanic island arcs that have formed where oceanic crust is subducting beneath another oceanic plate—that is, ocean-ocean plate boundaries. Andesitic stratovolcanoes can also form on continental margins, where oceanic crust is subducting under continental crust at an ocean-continent convergent plate boundary. The chains of stratovolcanoes of the Andes Mountains and the Cascades Range are examples of **continental volcanic arcs.**

What is a Wilson cycle?

____

____

Answer: The cycle of production of new oceanic crust, opening of the ocean basin, lateral movement of oceanic crust, and eventual consumption of the crust through subduction.

VOLCANIC ARC. These three andesitic stratovolcanoes on Russia's Kamchatka Peninsula—Klyuchevskoy (foreground), Kamen, and Bezymianny (rear)—are part of a volcanic arc formed as a result of the subduction of the Pacific Plate beneath Asia. Klyuchevskoy is seen here erupting volcanic ash. (Courtesy Brian Skinner)

7 IGNEOUS ROCKS IN CONTINENTAL TECTONIC ENVIRONMENTS

Ultimately, if plate convergence continues, the ocean basin trapped between converging plates may be completely consumed by subduction. Sometimes a small fragment of oceanic lithosphere—an ophiolite—is squeezed up and preserved along the plate boundary. Eventually, the continents on either shore of the ocean basin may collide to form great orogenic belts. Granitic magmas form at the base of continental crust that has been thickened as a result of a plate tectonic collision in an orogen. Unlike MORB, granites vary widely in composition, and that fact alone makes it unwise to generalize too much about the formation of granite. However, we know that granitic magma forms by wet partial melting of continental crust. The resulting magma is very rich in silica and extremely viscous. In fact, granitic magma is so viscous and moves upward so slowly that it mostly cools and crystallizes as great batholiths of **orogenic granite.** The small fraction of granitic magma that does reach the surface erupts as silica-rich rhyolitic lavas and tuffs.

How does granitic magma form, in general?

Answer: By wet partial melting of continental crust.

A few igneous rocks form in the interiors of plates, far removed from the intense activities at plate margins, in an **intraplate** environment. For example, consider what happens when a plate capped by continental crust passes over a plume-related mantle hot spot like the one that feeds the Hawaiian volcanoes. The rising magma must force its way through the continental crust, and the result is a vast outpouring of very fluid basaltic lava onto the continent. The lava forms a flat sheet of volcanic rock called a plateau basalt (also known as flood basalt, discussed in chapter 5). Plateau basalts, like oceanic shield volcanoes, are the products of plume-related decompression melting. The Columbia River basalts of Oregon and Washington were formed in this fashion, as were the plateau basalts of the Snake River region in Idaho. In fact, the Columbia and Snake River basalts came from the same plume-generated magma source. They are geographically separated from one another because the North American Plate is moving westward over the hot spot, just as the Pacific Plate is moving over the Hawaiian hot spot.

The mantle plume that fed the Columbia and Snake River plateau basalts is still active today; it now lies beneath Yellowstone National Park. Although there is a lot of basaltic rock at Yellowstone, much of the region is covered by a thick blanket of rhyolitic volcanic ash. The rhyolitic magma is thought to have formed by wet partial melting of continental crust that was heated by the rising hot basaltic magma. Some rhyolitic magma of the kind erupted at Yellowstone doesn't make it all the way to the surface. It remains in the crust and crystallizes to form bodies of granite. When such granites are exposed by erosion, it is clear that they have not formed in an orogen as a result of subduction or collision, like the orogenic granites discussed above. They are therefore called **anorogenic** ("not orogenic") **granites.**

Why do you think there are so many thermal springs and geysers at Yellowstone National Park?

Answer: The ground and the groundwater are heated by magma from the active hot spot that underlies the park.

Magmatic activity in continental plate interiors also occurs in association with rifts, which form when the lithosphere is stretched and thinned. Most continental rifts are spreading centers that have failed to develop (or haven't yet developed) into new oceans. Examples are the East African Rift, the Newark Basin of New

Jersey, and the Rio Grande Valley of New Mexico. In response to the thinning of the lithosphere in continental rift zones, the asthenosphere moves up. Decompression melting creates basaltic magma, which rises to form basaltic lava flows and gabbro sills and dikes.

Other interesting igneous rocks also form in continental rift environments. Among the most unusual are carbonatites, which are composed largely or entirely of the carbonate minerals calcite and dolomite. For many years it was not certain that carbonatites were igneous rocks. The rock textures were recognized as igneous, but it's hard to imagine where a magma with such an extreme composition—rich in carbonate but totally lacking in silica—could come from. Then, in 1962, carbonatite magma was observed erupting from a volcano in the East African Rift. The exact process by which carbonatite magmas form is still unclear, but it is reasonably certain that they originate in the upper mantle and that the degree of partial melting must be very small. Other rift-related magmas can have equally extreme compositions, including highly alkalic magmas with compositions similar to that of baking soda.

Kimberlite is another rare igneous rock, often associated with carbonatite. Kimberlites are poor in silica but rich in potassium—an unusual composition. Kimberlites, like carbonatites, are thought to form by very small degrees of partial melting of mantle rocks beneath old cratons. Kimberlites have the distinction of being the primary source for diamonds. Kimberlite magma apparently forms very deep in the mantle, as discussed in previous chapters. The magma contains a lot of dissolved carbon dioxide and it is intruded so forcefully that it punches a pipelike hole through the crust, like a cork coming out of a bottle of champagne. Kimberlite pipes, also called diamond pipes, range in diameter from a few meters to a few hundred meters.

What are carbonatites, and in what tectonic environment do they occur?

Answer: Carbonatites are igneous rocks formed almost entirely of calcite or dolomite; they occur in continental rift zones.

8 WHERE DIFFERENT METAMORPHIC ROCKS FORM AND WHY

As discussed in chapter 8, metamorphism is the term used to describe changes in mineral assemblage and rock texture induced in solid rocks as a result of increased temperature and pressure. The rocks subjected to change may be sedimentary, igneous, or even previously formed metamorphic rocks. The changes all take place in the solid state—that is, at temperatures and pressures below the onset of melting.

Recall that there are three basic kinds of metamorphism:

1. Contact metamorphism (changes induced in cool rocks that are intruded by a mass of hot magma)
2. Burial metamorphism (changes that result from increasing temperature and pressure in a deepening pile of sedimentary or volcanic rocks)
3. Regional metamorphism (changes induced in rock masses when they are squeezed and heated as a result of a plate collisions)

Contact metamorphism can occur in any situation where hot rocks come into contact with cool rocks, but burial and regional metamorphism occur in specific tectonic environments.

Burial metamorphism is most likely to occur along the **passive margin,** or trailing edge, of a continent (Figure 12.10). The Atlantic Coast of North America is a present-day example of a passive margin. The North American Plate is moving west-northwest, carrying the continent along as it moves away from the spreading plate margin at the Mid-Atlantic Ridge. Along passive continental margins like this, thick piles of sediment accumulate just offshore on the continental shelf and slope. Burial metamorphism occurs in the lower portions of the thick sequences of accumulated sediment. Wells drilled to depths of 5 km (3 mi) or more, seeking oil in the great pile of sediment accumulated off the mouth of the Mississippi River, have encountered extensive volumes of sedimentary rock altered by burial metamorphism.

Regional metamorphism is most likely to occur along the **active margin** of a continent, where tectonic, seismic, and volcanic activity are common. The

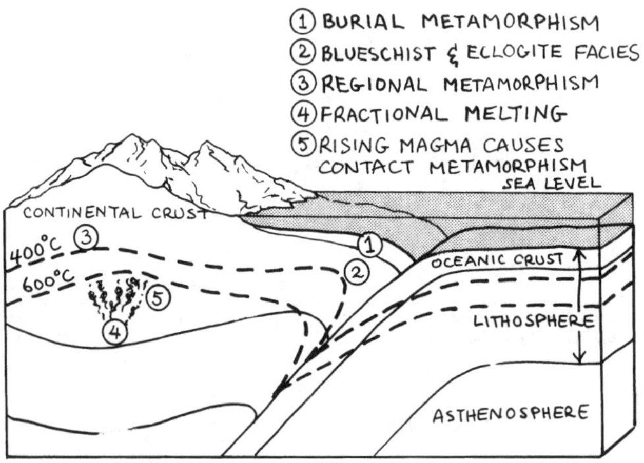

WHERE METAMORPHISM OCCURS

Figure 12.10

two most important tectonic settings for regional metamorphism are subduction zones at ocean-continent plate boundaries and collision zones along continent-continent plate boundaries. Let's look more closely at each of these.

When rocks undergo subduction, they are subjected to high temperatures and pressures. But changing the actual temperature and pressure of a rock is more complex than it might seem.

If you squeeze a cube of rock in a vise, all parts of the rock are immediately subjected to increased pressure. If you were to heat one side of the rock with a flame, it would be a long time before a thermometer on the other side of the rock registered a rise in temperature. The reason for the slow rise in temperature is that heat moves through a solid, rigid rock by conduction, and conduction is a slow process.

This is what happens during subduction, too—the pressure on the rock increases more quickly than the temperature. When a slab of cold lithosphere sinks into the hot asthenosphere, the pressure on the sinking lithosphere is equal to the pressure of the enclosing asthenosphere. However, because heat can only enter the slab by conduction, the sinking lithosphere heats up slowly. The combination of low temperature and high pressure in subduction zones produces distinctive conditions of metamorphism and distinctive kinds of metamorphic rocks (see Figure 12.10). At intermediate depths, a bluish-colored amphibole mineral (glaucophane) is produced in low-temperature, high-pressure metamorphic rocks called blueschists. Blueschists can be seen in many parts of the Coast Ranges of California, part of an old subduction zone. At greater depths, and therefore higher pressure, blueschists are replaced by eclogites, which contain a beautiful ruby-colored garnet and a green pyroxene mineral called jadeite ("jade"). Eclogites and jadeite are rare; they are found at the surface only where some tectonic event such as a plate collision has caused a fragment of subducting lithosphere to be brought back to the surface.

What is the difference between the tectonic conditions in which blueschists and eclogites are formed?

Answer: Both are formed in subduction zones—blueschists at intermediate depth (low temperature, high pressure), ecologites at greater depth (higher pressure).

When two continental masses collide, the result is a major mountain belt, such as the Himalayas or the Alps. The mountains stand high because the light continental crust is thickened by the collision process (see Figure 12.10). Because of isostasy there is a deep root beneath every mountain chain. Just as it does during subduction, the pressure rises more rapidly than the temperature in a thickening mass of crust. However, there is a limit to the thickening process and after some millions of years, the temperature finally catches up with the pressure. The result is the formation of regionally metamorphosed rocks—commonly slates, schists, and gneisses—all of which have distinctive fabrics and foliations that result from the squeezing associated with the collision process.

If the collision process that forms the mountain belt, or orogen, thickens the crust so much that partial melting starts at its base, orogenic granitic magma will be the result (see Figure 12.10). The magma rises and intrudes the overlying regionally metamorphosed rocks, forming batholiths. The existence of ancient orogens of regionally metamorphosed rocks intruded by granite batholiths provides some of the most convincing evidence that plate tectonics has been operating for a very long time.

Why are orogenic granites often associated with regionally metamorphosed rocks?

____

____

Answer: Both types of rock form deep in the roots of mountains, where continental crust has been thickened by continental collision and orogenesis.

9 WHERE DIFFERENT SEDIMENTARY ROCKS FORM AND WHY

Recall from chapter 7 that there are three kinds of sediment—clastic, chemical, and biogenic. Chemical and biogenic sediments form most commonly in shallow lakes and seas, where water evaporates or organic material accumulates. For example, an ocean basin trapped between converging plates may grow shallower and shallower. Eventually, the water may dry up altogether, leaving behind evaporite deposits.

Clastic sediments are far more abundant than chemical or biogenic sediments, and the locations where they form and accumulate are largely controlled by plate tectonics. Clastic sediment consists of loose, fragmented rock and mineral debris produced by weathering of continental rocks. The sediment is transported by erosion, eventually accumulating in low-lying areas. Rates of weathering and erosion are lowest in regions of low relief and highest in regions of active tectonics, where topographic relief is greatest—that is, in present-day mountain belts.

A variety of low-lying areas form within and along the edges of high mountain ranges, such as structural basins or troughs caused by the faulting and folding associated with mountain building. Coarse stream sediments eroded from a rising mountain range will accumulate in these low areas as vast thicknesses of conglomerates. The margins of the Himalayas in south-central Asia provide an example. Here, thick sequences of conglomerates and coarse sandstones flank the southern edge of the range, while the finest-grained sediments have been transported all the way to the sea by the many streams that flow to the Indian Ocean.

In continental rift valleys and long passive continental margins, thick wedges of clastic sediment slowly accumulate as streams transport sediment to the newly forming or existing ocean basin. Sediment eroded from the adjacent continent accumulates on a passive continental margin over many millions of years, and may reach thicknesses in excess of 10 km (6 mi). The Atlantic Ocean margin of North America is an example of such a site of accumulation. Most of the strata deposited on the shelf of a passive continental margin consist of shallow-water marine sediments. The continental shelf slowly subsides as accumulation takes place, and the pile of sediment grows thicker and thicker. At the elevated temperatures and pressures deep within the pile, diagenesis and burial metamorphism (the processes that collectively turn sediment into sedimentary rock) may occur.

The lowest-lying points on the Earth's surface are the long, linear, deep-ocean-floor trenches that are the topographic expression of subduction zones. Where subduction zones form near continental margins, as along the western margin of South America, the greatest topographic relief on the Earth is observed—from the bottom of the trench to the top of the volcanoes on the edge of the continent. Sediment is transported from the continent by streams and turbidity currents and accumulates in the deep trench. The presence of volcanoes on the land ensures that a lot of volcanic debris is present in the sediment. These wedge-shaped accumulations of clastic sediment in deep-ocean trenches are called **accretionary wedges.**

Some of the water-soaked sediment in deep-ocean trenches may be subducted into the mantle along with the sinking slab of oceanic lithosphere, releasing water that facilitates the wet melting of adjacent mantle. The rest of the sediment and volcanic debris is scraped off the subducting slab, ending up in piles of chaotically folded, jumbled, and broken-up material on the ocean floor. These scraped-off collections of sediment and volcanic debris, characteristic of the deep-ocean trench and subduction zone environment, are called **mélanges** (French for "mixtures").

Two other low-lying topographic features are usually associated with subduction zone trenches. A linear, troughlike depression often forms between the trench and the volcanic arc; this depression is called a **fore-arc basin.** Tensional forces created by the pull of the downgoing slab can also cause a depression to form behind the volcanic arc—a **back-arc basin.** Fore-arc and back-arc basins, while not as dramatic topographically as the deep-ocean trenches, are also low-lying areas where ocean-floor sediments accumulate (see Figure 12.9).

What is a mélange?

___

___

Answer: A chaotically folded, jumbled, and broken-up mixture of sedimentary and volcanic rock, scraped off the top of a downgoing slab of lithosphere in a subduction zone.

10 EARTH, OUR HOME

The theory of plate tectonics has revolutionized our understanding of the Earth. It has given us a context for the processes we observe in the rock cycle and the hydrologic cycle, and a mechanism through which we can link them to the inter-

EARTHRISE OVER THE MOON. The drab, barren, lifeless surface of the Moon in the foreground is in stark contrast to the atmosphere-enrobed and water-rich planet on which we live. (Courtesy NASA)

nal workings of the planet. What does the theory of plate tectonics explain? As you have just read, it explains where different types of rocks form and why. By the mid-1900s, geologists had accumulated great volumes of observations about rocks, their characteristics, and their locations. But until plate tectonics came along, there was no unified framework for these observations.

Plate tectonics also explains the distinctive topographic features of the Earth's surface, such as the deep-ocean trenches, high mountain chains, arc-shaped chains of volcanic islands, and long, linear faults. Geologists understand now that these and other topographic features are the surface expressions of distinctive internal Earth processes. We have even been able to use what we know about terrestrial tectonics to interpret the tectonic significance of many topographic features that have been observed on other planets, such as Mars and Venus; some of these features are earthlike, and others are very different indeed.

This book has introduced you to some of the fascinating processes that make the planet Earth the special place it is, and the wonderfully balanced ways by which it works and has worked over the vast geologic ages. But there is much about the Earth that we still know imperfectly or don't know at all. We continue to make discoveries about plate tectonics and how tectonic processes rearrange and renew the Earth's surface; about rainfall, winds, ice caps, and all the other forces that continually weather and erode the rocks that have been brought to the surface by volcanism and tectonic uplift; about climate and the global interactions that drive climatic change; about the influence of life on geologic processes, and vice versa; and about how humans can survive and flourish on a geologically active planet. I hope this book has created in you a new sense of enthusiasm and a commitment to continue learning about Earth, our home.

SELF-TEST

These questions are designed to help you assess how well you have learned the concepts presented in chapter 12. The answers are given at the end.

1. Which one of the following does not belong with the others?
   a. Hawaii
   b. mantle plume
   c. subduction
   d. Olympus Mons
2. Fragments of MORB that are found on land are ___________.
   a. remnants of ancient subduction zone volcanoes
   b. part of a distinctive group of rocks called ophiolites
   c. pieces of the mantle
   d. All of the above are true.

3. Which one of the following does not belong with the others?
   a. continental rift
   b. plateau basalt
   c. Columbia River and Snake River
   d. flood basalt

4. The flotational balance of the lithosphere on the asthenosphere is called ___________.

5. Magma that forms from the wet partial melting of subducting lithosphere and adjacent (overlying) mantle is ___________ in composition.

6. The two basic structural units that comprise the continents are ___________ and ___________.

7. The crust that makes up the floor of the oceans (away from spreading centers) is very old—as old as 3.8 billion years. (T or F)

8. When rocks are buried, the increase in pressure is transmitted to the rock more rapidly than the increase in temperature. (T or F)

9. Carbonatites are sedimentary rocks that form by precipitation of calcite and/or dolomite in a shallow marine environment. (T or F)

10. What is the significance of "pillowlike" structures in basaltic lavas?

    ____

11. What is the difference between orogenic and anorogenic granites?

    ____

    ____

    ____

12. What is the difference between a fore-arc basin and a back-arc basin?

    ____

    ____

    ____

13. What are the parts of an ophiolite, from the bottom up?

    ____

    ____

    ____

ANSWERS

1. c
2. b
3. a
4. isostasy
5. andesitic
6. cratons; orogens
7. F
8. T
9. F
10. Pillows indicate that the lava was extruded under water.
11. Orogenic granites occur in huge batholiths that form as a result of wet partial melting at the base of the continental crust, where the crust has been thickened as a result of a plate tectonic collision—that is, in an orogen. Anorogenic granites, in contrast, form in areas of the crust not related to an orogen. Some anorogenic granites may form as a result of partial melting at the base of the continental crust, when the continental plate passes over a mantle hot spot, as in Yellowstone Park.
12. A fore-arc basin is a troughlike depression that forms between a subduction zone trench and the volcanic island arc. A back-arc basin is a depression that forms behind the island arc, as a result of tensional forces created by the pull of the sinking slab.
13. Peridotite (upper mantle material); dikes and sills of gabbro; pillowed basalt (MORB); and metamorphosed seafloor sediments.

KEY WORDS

accretionary wedge
active margin
anorogenic granite
back-arc basin
closed system
continental shield
continental volcanic arc
craton
cycle
fore-arc basin
hot spot
hydrologic cycle
intraplate
isolated system
mantle plume
mélange
MORB (midocean ridge basalt)
open system
ophiolite
orogen
orogenic granite
passive margin
rock cycle
spilite
system
tectonic cycle
volcanic island arc
Wilson cycle

Appendix 1: Units and Conversions

ABOUT SI UNITS

Regardless of the field of specialization, all scientists use the same units and scales of measurement. They do so to avoid confusion and the possibility that mistakes can creep in when data are converted from one system of units, or one scale, to another. By international agreement the SI units are used by all, and they are the units used in this text. SI is the abbreviation of Système International d'Unités (in English, the International System of Units). Some of the SI units are likely to be familiar, some unfamiliar. The SI unit of length is the meter (m), of area the square meter (m^2), and of volume the cubic meter (m^3). The SI unit of mass is the kilogram (kg), and of time the second (s). The other SI units used in this book can be defined in terms of these basic units.

PREFIXES FOR VERY LARGE AND VERY SMALL NUMBERS

When very large or very small numbers have to be expressed, a standard set of prefixes is used in conjunction with the SI units. Some prefixes are probably already familiar; an example is the centimeter, which is one-hundredth of a meter, or 10^2 m). The standard prefixes are:

tera $1,000,000,000,000 = 10^{12}$
giga $1,000,000,000 = 10^9$
mega $1,000,000 = 10^6$
kilo $1,000 = 10^3$
hecto $100 = 10^2$
deka $10 = 10$
deci $0.1 = 10^{-1}$
centi $0.01 = 10^{-2}$
milli $0.001 = 10^{-3}$
micro $0.000001 = 10^{-6}$
nano $0.000000001 = 10^{-9}$
pico $0.000000000001 = 10^{-12}$

COMMONLY USED UNITS OF MEASURE

Length

Metric Measure

1 kilometer (km) = 1,000 meters (m)

1 meter (m) = 100 centimeters (cm)

1 centimeter (cm) = 10 millimeters (mm)

1 millimeter (mm) = 1,000 micrometers (μm) (formerly called microns)

1 micrometer (μm) = 0.001 millimeter (mm)

1 angstrom (Å) = 10^{-8} centimeters (cm)

Nonmetric Measure

1 mile (mi) = 5,280 feet (ft) = 1,760 yards (yd)

1 yard (yd) = 3 feet (ft)

1 foot (ft) = 12 inches (in)

1 fathom (fath) = 6 feet (ft)

Conversions

1 kilometer (km) = 0.6214 mile (mi)

1 meter (m) = 1.094 yards (yd) = 3.281 feet (ft)

1 centimeter (cm) = 0.3937 inch (in)

1 millimeter (mm) = 0.0394 inch (in)

1 mile (mi) = 1.609 kilometers (km)

1 yard (yd) = 0.9144 meter (m)

1 foot (ft) = 0.3048 meter (m)

1 inch (in) = 2.54 centimeters (cm)

1 inch (in) = 25.4 millimeters (mm)

1 fathom (fath) = 1.8288 meters (m)

Area

Metric Measure

1 square kilometer (km^2) = 1,000,000 square meters (m^2) = 100 hectares (ha)

1 square meter (m^2) = 10,000 square centimeters (cm^2)

1 hectare (ha) = 10,000 square meters (m^2)

Nonmetric Measure

1 square mile (mi^2) = 640 acres (ac)

1 acre (ac) = 4,840 square yards (yd^2)

1 square foot (ft^2) = 144 square inches (in^2)

Conversions

1 square kilometer (km^2) = 0.386 square mile (mi^2)

1 hectare (ha) = 2.471 acres (ac)

1 square meter (m^2) = 1.196 square yards (yd^2) = 10.764 square feet (ft^2)

1 square centimeter (cm^2) = 0.155 square inch (in^2)

1 square mile (mi^2) = 2.59 square kilometers (km^2)

1 acre (ac) = 0.4047 hectare (ha)

1 square yard (yd^2) = 0.836 square meter (m^2)

1 square foot (ft^2) = 0.0929 square meter (m^2)

1 square inch (in^2) = 6.4516 square centimeters (cm^2)

Volume

Metric Measure

1 cubic meter (m^3) = 1,000,000 cubic centimeters (cm^3)

1 liter (l) = 1,000 milliliters (ml) = 0.001 cubic meter (m^3)

1 centiliter (cl) = 10 milliliters (ml)

1 milliliter (ml) = 1 cubic centimeter (cm^3)

Nonmetric Measure

1 cubic yard (yd^3) = 27 cubic feet (ft^3)

1 cubic foot (ft^3) = 1,728 cubic inches (in^3)

1 barrel (oil) (bbl) = 42 gallons (U.S.) (gal)

Conversions

1 cubic kilometer (km^3) = 0.24 cubic mile (mi^3)

1 cubic meter (m^3) = 264.2 gallons (U.S.) (gal) = 35.314 cubic feet (ft^3)

1 liter (l) = 1.057 quarts (U.S.) (qt) = 33.815 ounces (U.S. fluid) (fl. oz.)

1 cubic centimeter (cm^3) = 0.0610 cubic inch (in^3)

1 cubic mile (mi^3) = 4.168 cubic kilometers (km^3)

1 acre-foot (ac-ft) = 1,233.46 cubic meters (m^3)

1 cubic yard (yd^3) = 0.7646 cubic meter (m^3)

1 cubic foot (ft^3) = 0.0283 cubic meter (m^3)

1 cubic inch (in^3) = 16.39 cubic centimeters (cm^3)

1 gallon (gal) = 3.784 liters (l)

Mass

Metric Measure

1,000 kilograms (kg) = 1 metric ton (also called a tonne) (m.t)

1 kilogram (kg) = 1,000 grams (g)

1 gram (g) = 0.001 kilogram (kg)

Nonmetric Measure

1 short ton (sh.t) = 2,000 pounds (lb)

1 long ton (l.t) = 2,240 pounds (lb)

1 pound (avoirdupois) (lb) = 16 ounces (avoirdupois) (oz) = 7,000 grains (gr)

1 ounce (avoirdupois) (oz) = 437.5 grains (gr)

1 pound (Troy) (Tr. lb) = 12 ounces (Troy) (Tr. oz)

1 ounce (Troy) (Tr. oz) = 20 pennyweight (dwt)

Conversions

1 metric ton (m.t) = 2,205 pounds (avoirdupois) (lb)

1 kilogram (kg) = 2.205 pounds (avoirdupois) (lb)

1 gram (g) = 0.03527 ounce (avoirdupois) (oz) = 0.03215 ounce (Troy) (Tr. oz) = 15,432 grains (gr)

1 pound (lb) = 0.4536 kilogram (kg)

1 ounce (avoirdupois) (oz) = 28.35 grams (g)

1 ounce (avoirdupois) (oz) = 1.097 ounces (Troy) (Tr. oz)

Pressure

1 pascal (Pa) = 1 newton/square meter (N/m^2)

1 kilogram/square centimeter (kg/cm^2) = 0.96784 atmosphere (atm) = 14.2233 pounds/square inch (lb/in^2) = 0.98067 bar

1 bar = 0.98692 atmosphere (atm) = 105 pascals (Pa) = 1.02 kilograms/square centimeter (kg/cm^2)

Energy and Power

Energy

1 joule (J) = 1 Newton meter (N.m) = 2.390×10^{-1} calorie (cal) = 9.47×10^{-4} British thermal unit (Btu) = 2.78×10^{-7} kilowatt-hour (kWh)

1 calorie (cal) = 4.184 joule (J) = 3.968×10^{-3} British thermal unit (Btu) = 1.16×10^{-6} kilowatt-hour (kWh)

1 British thermal unit (Btu) = 1,055.87 joules (J) = 252.19 calories (cal) = 2.928×10^{-4} kilowatt-hour (kWh)

1 kilowatt hour = 3.6×10^{6} joules (J) = 8.60×10^{5} calories (cal) = 3.41×10^{3} British thermal units (Btu)

Power (energy per unit time)

1 watt (W) = 1 joule per second (J/s) = 3.4129 British thermal units per hour (Btu/h) = 1.341×10^{-3} horsepower (hp) = 14.34 calories per minute (cal/min)

1 horsepower (hp) = 7.46×10^{2} watts (W)

Temperature

To change from Fahrenheit (F) to Celsius (C): °C = (°F − 32°C) ÷ 1.8

To change from Celsius (C) to Fahrenheit (F): °F = (°C × 1.8) + 32

To change from Celsius (C) to Kelvin (K): K = °C + 273.15

Appendix 2: Elements and Their Symbols

Elements and Their Symbols

Atomic Number	Name	Symbol	Atomic Number	Name	Symbol
1	Hydrogen	H	35	Bromine	Br
2	Helium	He	36	Krypton	Kr
3	Lithium	Li	37	Rubidium	Rb
4	Beryllium	Be	38	Strontium	Sr
5	Boron	B	39	Yttrium	Y
6	Carbon	C	40	Zirconium	Zr
7	Nitrogen	N	41	Niobium	Nb
8	Oxygen	O	42	Molybdenum	Mo
9	Fluorine	F	43	Technetium*	Tc
10	Neon	Ne	44	Ruthenium	Ru
11	Sodium	Na	45	Rhodium	Rh
12	Magnesium	Mg	46	Palladium	Pd
13	Aluminum	Al	47	Silver	Ag
14	Silicon	Si	48	Cadmium	Cd
15	Phosphorus	P	49	Indium	In
16	Sulfur	S	50	Tin	Sn
17	Chlorine	Cl	51	Antimony	Sb
18	Argon	Ar	52	Tellurium	Te
19	Potassium	K	53	Iodine	I
20	Calcium	Ca	54	Xenon	Xe
21	Scandium	Sc	55	Cesium	Cs
22	Titanium	Ti	56	Barium	Ba
23	Vanadium	V	57	Lanthanum	La
24	Chromium	Cr	58	Cerium	Ce
25	Manganese	Mn	59	Praseodymium	Pr
26	Iron	Fe	60	Neodymium	Nd
27	Cobalt	Co	61	Promethium*	Pm
28	Nickel	Ni	62	Samarium	Sm
29	Copper	Cu	63	Europium	Eu
30	Zinc	Zn	64	Gadolinium	Gd
31	Gallium	Ga	65	Terbium	Tb
32	Germanium	Ge	66	Dysprosium	Dy
33	Arsenic	As	67	Holmium	Ho
34	Selenium	Se	68	Erbium	Er

(*Continued*)

Elements and Their Symbols (*Continued*)

Atomic Number	Name	Symbol	Atomic Number	Name	Symbol
69	Thulium	Tm	90	Thorium	Th
70	Ytterbium	Yb	91	Protactinium	Pa
71	Lutetium	Lu	92	Uranium	U
72	Hafnium	Hf	93	Neptunium*	Np
73	Tantalum	Ta	94	Plutonium*	Pu
74	Tungsten	W	95	Americium*	Am
75	Rhenium	Re	96	Curium*	Cm
76	Osmium	Os	97	Berkelium*	Bk
77	Iridium	Ir	98	Californium*	Cf
78	Platinum	Pt	99	Einsteinium*	Es
79	Gold	Au	100	Fermium*	Fm
80	Mercury	Hg	101	Mendelevium*	Md
81	Thallium	Tl	102	Nobelium*	No
82	Lead	Pb	103	Lawrencium*	Lr
83	Bismuth	Bi	104	Unnilquadium*	Unq
84	Polonium	Po	105	Unnilpentium*	Unp
85	Astatine*	At	106	Unnilhexium*	Unh
86	Radon	Rn	107	Unnilseptium*	Uns
87	Francium*	Fr	108	Unniloctium*	Uno
88	Radium	Ra	109	Unnilennium*	Une
89	Actinium*	Ac			

*These elements are humanly made and are not known to occur naturally.

Appendix 3: Properties of Some Important Minerals

Table A3.1 Silicate Minerals

Mineral Group	Compositions of Important Examples	Crystal Form and Habit	Cleavage	Hardness	Color and Luster	Other Properties
Feldspars	Plagioclase $(Na,Ca)(Si,Al)_3O_8$	Framework silicate; irregular grains, tabular crystals	Two, perfect, not quite at right angles	6–6.5	White to dark gray	Fine parallel striations (twins) on cleavage planes; common mineral
	Alkali feldspars $(K,Na)(Si_3Al)O_8$	Framework silicate; prism-shaped crystals	Two, perfect, at right angles	6	Flesh-colored, pink, white, or gray	Common mineral
Quartz	Quartz SiO_2	Framework silicate; 6-sided crystals	None	7	Colorless, white, gray; vitreous luster	Common mineral; may have other colors caused by impurities (e.g., amethyst, purple)
Micas	Biotite $K(Mg,Fe)_3(Si_3Al)O_{10}(OH)_2$	Sheet silicate; flaky, platy	One, perfect	2.5–3	Black, brown, dark green	Common in igneous and metamorphic rocks
	Muscovite $KAl(Si_3Al)O_{10}(OH)_2$	Sheet silicate; flaky, platy	One, perfect	2–2.5	Colorless, pale green, pale brown	Common in igneous and metamorphic rocks

Pyroxenes	Augite $Ca(Mg,Fe,Al)[(Si,Al)O_3]_2$	Chain silicate; 8-sided stubby crystals	Two, nearly at right angles	5–6	Dark green to black	Common in igneous rocks; other varieties may be white to green
Olivines	Olivine $(Mg,Fe)_2SiO_4$	Isolated silica tetrahedra; small grains, granular masses	None	6.5–7	Olive green to yellowish green	Common in igneous rocks
Chlorites	Chlorite $(Mg,Fe,Al)_6(Si,Al)_4O_{10}(OH)_8$	Sheet silicate; flaky masses of minute scales	One, perfect	2–2.5	Light to dark green; greasy luster	Common in metamorphic rocks
Amphiboles	Hornblende $(Ca,Na)_2(Mg,Fe,Al)_5Si_8O_{22}(OH_2)$	Chain silicate; long, 6-sided crystals; also fibers	Two, intersecting at 56° and 124°	5–6	Green to black; some varieties white	Common in metamorphic rocks
Garnets	Garnet $(Ca,Mg,Fe,Mn)_3(Al,Fe,Ti,Cr)_2(SiO_4)_3$	Isolated silica tetrahedra; 12- or 24-sided crystals	None	6.5–7.5	Red, brown, yellowish green, black	Common in metamorphic rocks
Clays	Kaolinite $Al_4Si_4O_{10}(OH)_8$	Sheet silicate; soft, earthy masses	One, perfect	2–2.5	White, yellowish; dull luster	Common in sediments, soils, and metamorphic rocks; plastic when wet; clayey odor

Table A3.2 Nonsilicate Minerals

Mineral Group	Compositions of Important Examples	Habit	Cleavage	Hardness	Color and Luster	Other Properties
Carbonates	Calcite $CaCO_3$	Tapering, rhomb-shaped crystals and granular masses	Three, perfect	3	Colorless or white; pearly luster	Effervesces in dilute HCl
	Dolomite $CaMg(CO_3)_2$	Tapering, rhomb-shaped crystals and granular masses	Three, perfect	3.5	White or gray; pearly luster	Does not effervesce in HCl unless powdered
Oxides	Hematite Fe_2O_3	Massive, granular	None	5–6	Reddish brown to black; metallic luster	Reddish brown streak
	Magnetite Fe_3O_4	Massive, granular; octahedral crystals	None	5.5–6.5	Black; metallic luster	Black streak; strongly attracted to magnets
Sulfides	Pyrite ("fool's gold") FeS_2	Cubic crystals with striated faces	None	6–6.5	Pale brass-yellow; metallic luster	Greenish black streak; not malleable, which distinguishes it from gold
	Pyrrhotite FeS	Granular; crystals rare	None	4	Brownish bronze; metallic luster	Black streak; magnetic
Sulfates	Anhydrite $CaSO_4$	Crystals rare; irregular grains and fibers	Three, at right angles	3	White or colorless; pearly luster	Alters to gypsum
	Gypsum $CaSO_4 \cdot 2H_2O$	Elongate or tabular crystals; fibrous and earthy masses	One, perfect; flakes bend	2	Colorless; vitreous to pearly luster	Most common sulfate mineral
Phosphates	Apatite $Ca_5(PO_4)_3(F,OH)$	6-sided crystals	One, poor	5	Green, brown, blue, or white	Common in many rocks in small amounts

Appendix 4: Symbols Commonly Used on Geologic Maps

Symbols Commonly Used on Geologic Maps

Symbol	Explanation
	Strike and dip of strata
 90	Strike of vertical strata
	Horizontal strata, no strike, dip = 0°
 43	Strike and dip of foliation in metamorphic rocks
	Strike of vertical foliation
	Anticline; arrows show directions of dip away from axis
	Syncline; arrows show directions of dip toward axis
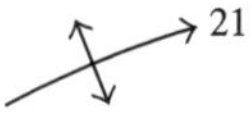 21	Anticline; showing direction and angle of plunge
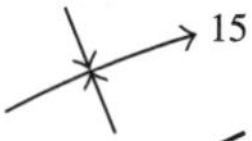 15	Syncline; showing direction and angle of plunge
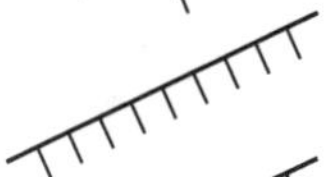	Normal fault; hachures on downthrown side
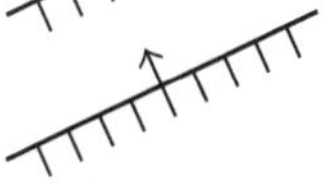	Reverse fault; arrow shows direction of dip, hachures on downthrown side
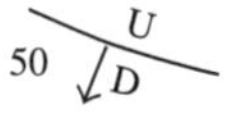 50 U D	Dip of fault surface; D, downthrown side; U, upthrown side
	Directions of relative horizontal movement along a fault
	Low-angle thrust fault; barbs on upper block

Map Symbols for Rock Types

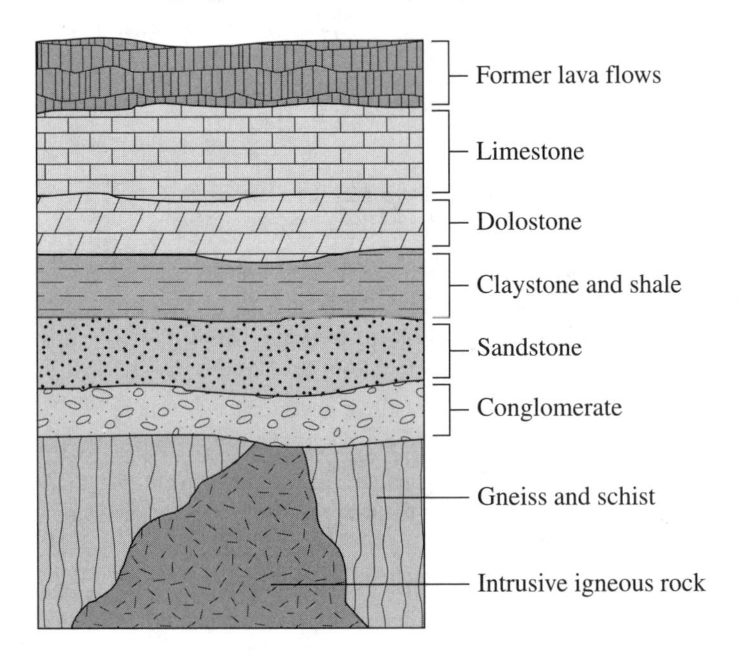

Appendix 5: Some Great Earth Science Web Sites

Here are just a few of the hundreds of great geology and Earth science web sites that provide free access to information, activities, and educational resources for both kids and adults.

For more information about . . .	try visiting . . .
Plate Tectonics	• www.platetectonics.com/Tharpe-Heezen, for lots of information about plate tectonics and for a close look at the famous Tharpe-Heezen map of the ocean floors.
What the Earth Is Made Of	• www.quiz.com/jq/12161list.html. Take the quizzes on igneous, metamorphic, and sedimentary rocks. • www.america-alfresco.com/gems/basicrocks.html, an extensive site for mineral collectors and anyone interested in minerals or gems. • www.minerals.net, The Mineral and Gemstone Kingdom, a free, interactive, educational guide to minerals and gemstones.
The Rock Record and Geologic Time	• www.ucmp.berkeley.edu/fosrec/fosrec.html, part of the Museum of Paleontology at the University of California, Berkeley, an exhibit entitled Learning from the Fossil Record.
Earthquakes and the Inside of the Earth	• www.crustal.ucsb.edu/ics/understanding. The University of California at Santa Barbara has extensive geology and paleontology web sites, including this one, Understanding Earthquakes.
Volcanoes and Igneous Rocks	• volcano.und.nodak.edu/, a truly fun, interactive site with lots of information, pictures, and up-to-date data on volcanoes around the world.
Weathering and Erosion	• pages.prodigy.net/pmedina/erosion.htm for definitions and examples of different types of erosion and weathering.
Sediments and Sedimentary Rocks	• www.fi.edu/fellows/payton/rocks/create/sediment.html for basic information and animations showing how sediments and sedimentary rocks (and other types of rocks) are created.

Metamorphism and Rock Deformation	• www.personal.umich.edu/~vdpluijm/earthstructure.html for extensive information, graphics, and animations concerning structural geology. • www.geolab.unc.edu/Petunia/IgMetAtlas/mainmenu.html for an atlas of rocks under the microscope.
The Hydrosphere and the Atmosphere	• www.nws.noaa.gov. The National Atmospheric and Oceanographic Administration has extensive offerings on just about everything to do with the atmosphere and hydrosphere, including real-time data. • www.nws.noaa.gov/om/, the web site of the National Weather Service Office of Meteorology.
The Record of Life on Earth	• www.ucmp.berkeley.edu, where you can wander virtually through the Museum of Paleontology at the University of California, Berkeley.
Resources from the Earth	• minerals.usgs.gov/, the Mineral Resources group from the U.S. Geological Survey, with lots of information about mineral resources and their value.
Earth Systems and Cycles (and General Geology)	• www.usgs.gov. The U.S. Geological Survey—the logical place to begin for all things geological. • www.usgs.gov/sci_challenge, part of the U.S. Geological Survey's web site, where you can take their 196-question "Science Challenge" test and see how well you do. • walrus.wr.usgs.gov/ask-a-geologist/. Submit your geological questions and a USGS geologist will answer them for you. • www-sci.lib.uci.edu/SEP/SEP.html. Frank Potter's Earth Science Gems, with links to thousands of great Earth science and other types of web sites. • agcwww.bio.ns.ca/schools/schools-index.html, Geological Survey of Canada educational resources. • www.jpl.nasa.gov/, the extensive set of web sites offered by the Jet Propulsion Lab and NASA, including just about anything you ever wanted to know about *other* planets.

Index

Page numbers in *italic* type indicate illustrations.